I0038380

Statistics and Sampling Theory

Statistics and Sampling Theory

Edited by
Casey Murphy

www.willfordpress.com

Published by Willford Press,
118-35 Queens Blvd., Suite 400,
Forest Hills, NY 11375, USA

ISBN: 978-1-68285-494-5

Cataloging-in-Publication Data

Statistics and sampling theory / edited by Casey Murphy.
p. cm.
Includes bibliographical references and index.
ISBN 978-1-68285-494-5
1. Sampling (Statistics). 2. Statistics. I. Murphy, Casey.

QA276.6 .S73 2018
519.52--dc23

For information on all Willford Press publications
visit our website at www.willfordpress.com

Contents

Preface

Statistics is an essential part of mathematics. It deals with the presentation, collection, organization, analysis and interpretation of data. In statistics, sampling theory refers to the survey method in which a particular group of people are chosen to represent the larger population. It is a very commonly used method in research and marketing. The different types of sampling methods are systematic sampling, cluster sampling, snowball sampling, minimax sampling, stratified sampling, theoretical sampling, etc. This textbook is a valuable compilation of topics, ranging from the basic to the most complex theories and principles in the field of statistics and sampling theory. It will serve as a valuable source of reference for those interested in this field.

A short introduction to every chapter is written below to provide an overview of the content of the book:

Chapter 1 - Sampling is the collection and interpretation of data. Some of the advantages of sampling are lower costs and the easier accumulation of data. This chapter is an overview of the subject matter incorporating all the major aspects of sampling; **Chapter 2 -** Simple random sampling is the collection of data taken from a large population. Programming and discrete uniform distribution are some of the methods by which the processes related to simple random sampling can be implemented. The chapter strategically encompasses and incorporates the major components and key concepts of simple random sampling, providing a complete understanding; **Chapter 3 -** Stratified sampling is the method of sampling that involves the division of the population. It categorizes the population into smaller groups which are known as strata. The groups are formed on the basis of similarity of attributes or characteristics. Stratified sampling is best understood in confluence with the major topics listed in the following chapter; **Chapter 4 -** Cluster sampling is seen when the subsets of a set are mutually homogenous but the elements of each set have heterogeneous characteristics in a statistical population. In this, the whole set is divided into small subsets called clusters and simple random sample method is applied. The chapter closely examines the key concepts of cluster sampling to provide an extensive understanding of the subject; **Chapter 5 -** Systematic sampling is a sampling technique that is methodical as the selection procedure follows a decided pattern. It is considered to be more convenient than random sampling method and the chances of equal probability of selection of elements remains same. The section serves as a source to understand the major categories related to systematic sampling. This chapter elucidates the crucial theories and principles of systematic sampling.

I extend my sincere thanks to the publisher for considering me worthy of this task. Finally, I thank my family for being a source of support and help.

Editor

1

Basics of Sampling

Sampling is the collection and interpretation of data. Some of the advantages of sampling are lower costs and the easier accumulation of data. This chapter is an overview of the subject matter incorporating all the major aspects of sampling.

Sampling (Statistics)

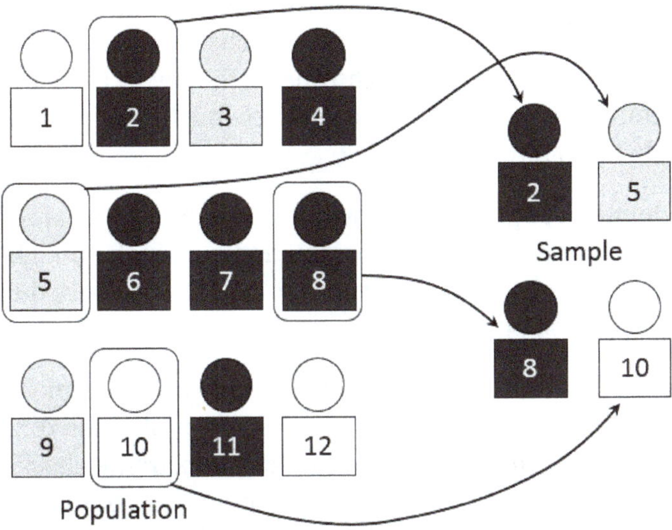

A visual representation of the sampling process.

In statistics, quality assurance, and survey methodology, sampling is concerned with the selection of a subset of individuals from within a statistical population to estimate characteristics of the whole population. Two advantages of sampling are that the cost is lower and data collection is faster than measuring the entire population.

Each observation measures one or more properties (such as weight, location, color) of observable bodies distinguished as independent objects or individuals. In survey sampling, weights can be applied to the data to adjust for the sample design, particularly stratified sampling. Results from probability theory and statistical theory are employed to guide the practice. In business and medical research, sampling is widely used for gathering information about a population. Acceptance sampling is used to determine if a production lot of material meets the governing specifications.

The sampling process comprises several stages:

- Defining the population of concern
- Specifying a sampling frame, a set of items or events possible to measure
- Specifying a sampling method for selecting items or events from the frame
- Determining the sample size
- Implementing the sampling plan
- Sampling and data collecting

Population Definition

Successful statistical practice is based on focused problem definition. In sampling, this includes defining the population from which our sample is drawn. A population can be defined as including all people or items with the characteristic one wishes to understand. Because there is very rarely enough time or money to gather information from everyone or everything in a population, the goal becomes finding a representative sample (or subset) of that population.

Sometimes what defines a population is obvious. For example, a manufacturer needs to decide whether a batch of material from production is of high enough quality to be released to the customer, or should be sentenced for scrap or rework due to poor quality. In this case, the batch is the population.

Although the population of interest often consists of physical objects, sometimes we need to sample over time, space, or some combination of these dimensions. For instance, an investigation of supermarket staffing could examine checkout line length at various times, or a study on endangered penguins might aim to understand their usage of various hunting grounds over time. For the time dimension, the focus may be on periods or discrete occasions.

In other cases, our 'population' may be even less tangible. For example, Joseph Jagger studied the behaviour of roulette wheels at a casino in Monte Carlo, and used this to identify a biased wheel. In this case, the 'population' Jagger wanted to investigate was the overall behaviour of the wheel (i.e. the probability distribution of its results over infinitely many trials), while his 'sample' was formed from observed results from that wheel. Similar considerations arise when taking repeated measurements of some physical characteristic such as the electrical conductivity of copper.

This situation often arises when we seek knowledge about the cause system of which the *observed* population is an outcome. In such cases, sampling theory may treat the observed population as a sample from a larger 'superpopulation'. For example, a researcher might study the success rate of a new 'quit smoking' program on a test group of 100 patients, in order to predict the effects of the program if it were made available nationwide. Here the superpopulation is "everybody in the country, given access to this treatment" – a group which does not yet exist, since the program isn't yet available to all.

Note also that the population from which the sample is drawn may not be the same as the population about which we actually want information. Often there is large but not complete overlap between these two groups due to frame issues etc. Sometimes they may be entirely separate – for instance, we might study rats in order to get a better understanding of human health, or we might study records from people born in 2008 in order to make predictions about people born in 2009.

Time spent in making the sampled population and population of concern precise is often well spent, because it raises many issues, ambiguities and questions that would otherwise have been overlooked at this stage.

Sampling Frame

In the most straightforward case, such as the sampling of a batch of material from production (acceptance sampling by lots), it would be most desirable to identify and measure every single item in the population and to include any one of them in our sample. However, in the more general case this is not usually possible or practical. There is no way to identify all rats in the set of all rats. Where voting is not compulsory, there is no way to identify which people will actually vote at a forthcoming election (in advance of the election). These imprecise populations are not amenable to sampling in any of the ways below and to which we could apply statistical theory.

As a remedy, we seek a sampling frame which has the property that we can identify every single element and include any in our sample. The most straightforward type of frame is a list of elements of the population (preferably the entire population) with appropriate contact information. For example, in an opinion poll, possible sampling frames include an electoral register and a telephone directory.

A probability sample is a sample in which every unit in the population has a chance (greater than zero) of being selected in the sample, and this probability can be accurately determined. The combination of these traits makes it possible to produce unbiased estimates of population totals, by weighting sampled units according to their probability of selection.

Example: We want to estimate the total income of adults living in a given street. We visit each household in that street, identify all adults living there, and randomly select one adult from each household. (For example, we can allocate each person a random number, generated from a uniform distribution between 0 and 1, and select the person with the highest number in each household). We then interview the selected person and find their income.

People living on their own are certain to be selected, so we simply add their income to our estimate of the total. But a person living in a household of two adults has only a one-in-two chance of selection. To reflect this, when we come to such a household, we would count the selected person's income twice towards the total. (The person who is selected from that household can be loosely viewed as also representing the person who isn't *selected).*

In the above example, not everybody has the same probability of selection; what makes it a probability sample is the fact that each person's probability is known. When every element in the population *does* have the same probability of selection, this is known as an 'equal probability of selection' (EPS) design. Such designs are also referred to as 'self-weighting' because all sampled units are given the same weight.

Probability sampling includes: Simple Random Sampling, Systematic Sampling, Stratified Sampling, Probability Proportional to Size Sampling, and Cluster or Multistage Sampling. These various ways of probability sampling have two things in common:

1. Every element has a known nonzero probability of being sampled and
2. Involves random selection at some point.

Nonprobability Sampling

Nonprobability sampling is any sampling method where some elements of the population have *no* chance of selection (these are sometimes referred to as 'out of coverage'/'undercovered'), or where the probability of selection can't be accurately determined. It involves the selection of elements based on assumptions regarding the population of interest, which forms the criteria for selection. Hence, because the selection of elements is nonrandom, nonprobability sampling does not allow the estimation of sampling errors. These conditions give rise to exclusion bias, placing limits on how much information a sample can provide about the population. Information about the relationship between sample and population is limited, making it difficult to extrapolate from the sample to the population.

Example: We visit every household in a given street, and interview the first person to answer the door. In any household with more than one occupant, this is a nonprobability sample, because some people are more likely to answer the door (e.g. an unemployed person who spends most of their time at home is more likely to answer than an employed housemate who might be at work when the interviewer calls) and it's not practical to calculate these probabilities.

Nonprobability sampling methods include convenience sampling, quota sampling and purposive sampling. In addition, nonresponse effects may turn *any* probability design into a nonprobability design if the characteristics of nonresponse are not well understood, since nonresponse effectively modifies each element's probability of being sampled.

Sampling Methods

Within any of the types of frame identified above, a variety of sampling methods can be employed, individually or in combination. Factors commonly influencing the choice between these designs include:

- Nature and quality of the frame
- Availability of auxiliary information about units on the frame
- Accuracy requirements, and the need to measure accuracy
- Whether detailed analysis of the sample is expected
- Cost/operational concerns

Simple Random Sampling

In a simple random sample (SRS) of a given size, all such subsets of the frame are given an equal probability. Furthermore, any given *pair* of elements has the same chance of selection as any other such pair (and similarly for triples, and so on). This minimizes bias and simplifies analysis of results. In particular, the variance between individual results within the sample is a good indicator of variance in the overall population, which makes it relatively easy to estimate the accuracy of results.

SRS can be vulnerable to sampling error because the randomness of the selection may result in a sample that doesn't reflect the makeup of the population. For instance, a simple random sample of ten people from a given country will *on average* produce five men and five women, but any given trial is likely to overrepresent one sex and underrepresent the other. Systematic and stratified techniques attempt to overcome this problem by "using information about the population" to choose a more "representative" sample.

SRS may also be cumbersome and tedious when sampling from an unusually large target population. In some cases, investigators are interested in "research questions specific" to subgroups of the population. For example, researchers might be interested in examining whether cognitive ability as a predictor of job performance is equally applicable across racial groups. SRS cannot accommodate the needs of researchers in this situation because it does not provide subsamples of the population. "Stratified sampling" addresses this weakness of SRS.

Systematic Sampling

Systematic sampling (also known as interval sampling) relies on arranging the study population according to some ordering scheme and then selecting elements at regular intervals through that ordered list. Systematic sampling involves a random start and then proceeds with the selection of every *k*th element from then onwards. In this case, *k*=(population size/sample size). It is important that the starting point is not automatically the first in the list, but is instead randomly chosen from within the first to the *k*th element in the list. A simple example would be to select every 10th name from the telephone directory (an 'every 10th' sample, also referred to as 'sampling with a skip of 10').

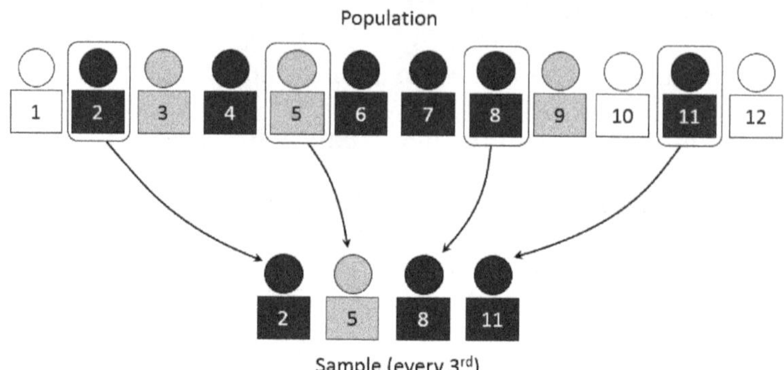

A visual representation of selecting a random sample using the systematic sampling technique

As long as the starting point is randomized, systematic sampling is a type of probability sampling. It is easy to implement and the stratification induced can make it efficient, *if* the variable by which the list is ordered is correlated with the variable of interest. 'Every 10th' sampling is especially useful for efficient sampling from databases.

For example, suppose we wish to sample people from a long street that starts in a poor area (house No. 1) and ends in an expensive district (house No. 1000). A simple random selection of addresses from this street could easily end up with too many from the high end and too few from the low end (or vice versa), leading to an unrepresentative sample. Selecting (e.g.) every 10th street number along the street ensures that the sample is spread evenly along the length of the street, representing all of these districts. (Note that if we always start at house #1 and end at #991, the sample is slightly biased towards the low end; by randomly selecting the start between #1 and #10, this bias is eliminated.

However, systematic sampling is especially vulnerable to periodicities in the list. If periodicity is present and the period is a multiple or factor of the interval used, the sample is especially likely to be *un*representative of the overall population, making the scheme less accurate than simple random sampling.

For example, consider a street where the odd-numbered houses are all on the north (expensive) side of the road, and the even-numbered houses are all on the south (cheap) side. Under the sampling scheme given above, it is impossible to get a representative sample; either the houses sampled will *all* be from the odd-numbered, expensive side, or they will *all* be from the even-numbered, cheap side, unless the researcher has previous knowledge of this bias and avoids it by a using a skip which ensures jumping between the two sides (any odd-numbered skip).

Another drawback of systematic sampling is that even in scenarios where it is more accurate than SRS, its theoretical properties make it difficult to *quantify* that accuracy. (In the two examples of systematic sampling that are given above, much of the potential sampling error is due to variation between neighbouring houses – but because this method never selects two neighbouring houses, the sample will not give us any information on that variation).

As described above, systematic sampling is an EPS method, because all elements have the same probability of selection (in the example given, one in ten). It is *not* 'simple random sampling' because different subsets of the same size have different selection probabilities – e.g. the set {4,14,24,...,994} has a one-in-ten probability of selection, but the set {4,13,24,34,...} has zero probability of selection.

Systematic sampling can also be adapted to a non-EPS approach; for an example.

Stratified Sampling

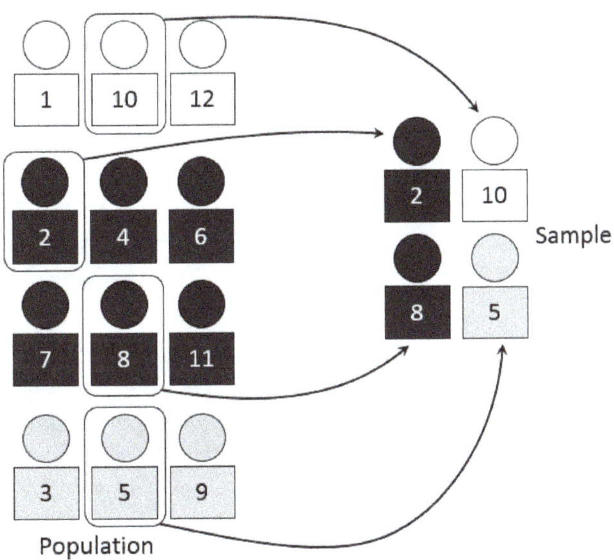

A visual representation of selecting a random sample using the stratified sampling technique

When the population embraces a number of distinct categories, the frame can be organized by these categories into separate "strata." Each stratum is then sampled as an independent sub-population, out of which individual elements can be randomly selected. There are several potential benefits to stratified sampling.

First, dividing the population into distinct, independent strata can enable researchers to draw inferences about specific subgroups that may be lost in a more generalized random sample.

Second, utilizing a stratified sampling method can lead to more efficient statistical estimates (provided that strata are selected based upon relevance to the criterion in question, instead of availability of the samples). Even if a stratified sampling approach does not lead to increased statistical efficiency, such a tactic will not result in less efficiency than would simple random sampling, provided that each stratum is proportional to the group's size in the population.

Third, it is sometimes the case that data are more readily available for individual, pre-existing strata within a population than for the overall population; in such cases, using a stratified sampling approach may be more convenient than aggregating data across groups (though this may potentially be at odds with the previously noted importance of utilizing criterion-relevant strata).

Finally, since each stratum is treated as an independent population, different sampling approaches can be applied to different strata, potentially enabling researchers to use the approach best suited (or most cost-effective) for each identified subgroup within the population.

There are, however, some potential drawbacks to using stratified sampling. First, identifying strata and implementing such an approach can increase the cost and complexity of sample selection, as well as leading to increased complexity of population estimates. Second, when examining multiple criteria, stratifying variables may be related to some, but not to others, further complicating the design, and potentially reducing the utility of the strata. Finally, in some cases (such as designs with a large number of strata, or those with a specified minimum sample size per group), stratified sampling can potentially require a larger sample than would other methods (although in most cases, the required sample size would be no larger than would be required for simple random sampling.

A stratified sampling approach is most effective when three conditions are met

1. Variability within strata are minimized
2. Variability between strata are maximized
3. The variables upon which the population is stratified are strongly correlated with the desired dependent variable.

Advantages over other sampling methods

1. Focuses on important subpopulations and ignores irrelevant ones.
2. Allows use of different sampling techniques for different subpopulations.
3. Improves the accuracy/efficiency of estimation.
4. Permits greater balancing of statistical power of tests of differences between strata by sampling equal numbers from strata varying widely in size.

Disadvantages

1. Requires selection of relevant stratification variables which can be difficult.
2. Is not useful when there are no homogeneous subgroups.
3. Can be expensive to implement.

Poststratification

Stratification is sometimes introduced after the sampling phase in a process called "post-stratification". This approach is typically implemented due to a lack of prior knowledge of an appropriate stratifying variable or when the experimenter lacks the necessary information to create a stratifying variable during the sampling phase. Although the method is susceptible to the pitfalls of post hoc approaches, it can provide several benefits in the right situation. Implementation usually follows a simple random sample. In addition to allowing for stratification on an ancillary variable, poststratification can be used to implement weighting, which can improve the precision of a sample's estimates.

Oversampling

Choice-based sampling is one of the stratified sampling strategies. In choice-based sampling, the data are stratified on the target and a sample is taken from each stratum so that the rare target class will be more represented in the sample. The model is then built on this biased sample. The effects of the input variables on the target are often estimated with more precision with the choice-based sample even when a smaller overall sample size is taken, compared to a random sample. The results usually must be adjusted to correct for the oversampling.

Probability-proportional-to-size Sampling

In some cases the sample designer has access to an "auxiliary variable" or "size measure", believed to be correlated to the variable of interest, for each element in the population. These data can be used to improve accuracy in sample design. One option is to use the auxiliary variable as a basis for stratification.

Another option is probability proportional to size ('PPS') sampling, in which the selection probability for each element is set to be proportional to its size measure, up to a maximum of 1. In a simple PPS design, these selection probabilities can then be used as the basis for Poisson sampling. However, this has the drawback of variable sample size, and different portions of the population may still be over- or under-represented due to chance variation in selections.

Systematic sampling theory can be used to create a probability proportionate to size sample. This is done by treating each count within the size variable as a single sampling unit. Samples are then identified by selecting at even intervals among these counts within the size variable. This method is sometimes called PPS-sequential or monetary unit sampling in the case of audits or forensic sampling.

Example: Suppose we have six schools with populations of 150, 180, 200, 220, 260, and 490 students respectively (total 1500 students), and we want to use student population as the basis for a PPS sample of size three. To do this, we could allocate the first school numbers 1 to 150, the second school 151 to 330 (= 150 + 180), the third school 331 to 530, and so on to the last school (1011 to 1500). We then generate a random start between 1 and 500 (equal to 1500/3) and count through the school populations by multiples of 500. If our random start was 137, we would select the schools which have been allocated numbers 137, 637, and 1137, i.e. the first, fourth, and sixth schools.

The PPS approach can improve accuracy for a given sample size by concentrating sample on large elements that have the greatest impact on population estimates. PPS sampling is commonly used for surveys of businesses, where element size varies greatly and auxiliary information is often available—for instance, a survey attempting to measure the number of guest-nights spent in hotels might use each hotel's number of rooms as an auxiliary variable. In some cases, an older measurement of the variable of interest can be used as an auxiliary variable when attempting to produce more current estimates.

Cluster Sampling

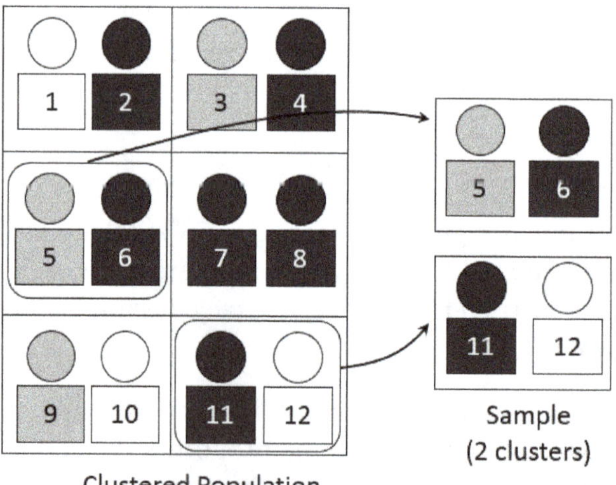

A visual representation of selecting a random sample using the cluster sampling technique

Sometimes it is more cost-effective to select respondents in groups ('clusters'). Sampling is often clustered by geography, or by time periods. (Nearly all samples are in some sense 'clustered' in time – although this is rarely taken into account in the analysis.) For instance, if surveying households within a city, we might choose to select 100 city blocks and then interview every household within the selected blocks.

Clustering can reduce travel and administrative costs. An interviewer can make a single trip to visit several households in one block, rather than having to drive to a different block for each household.

It also means that one does not need a sampling frame listing all elements in the target population. Instead, clusters can be chosen from a cluster-level frame, with an element-level frame created only for the selected clusters. The sample only requires a block-level city map for initial selections, and then a household-level map of the 100 selected blocks, rather than a household-level map of the whole city.

Cluster sampling (also known as clustered sampling) generally increases the variability of sample estimates above that of simple random sampling, depending on how the clusters differ between one another as compared to the within-cluster variation. For this reason, cluster sampling requires a larger sample than SRS to achieve the same level of accuracy – but cost savings from clustering might still make this a cheaper option.

Cluster sampling is commonly implemented as multistage sampling. This is a complex form of cluster sampling in which two or more levels of units are embedded one in the other. The first stage consists of constructing the clusters that will be used to sample from. In the second stage, a sample of primary units is randomly selected from each cluster (rather than using all units contained in all selected clusters). In following stages, in each of those selected clusters, additional samples of units are selected, and so on. All ultimate units (individuals, for instance) selected at the last step of this procedure are then surveyed. This technique, thus, is essentially the process of taking random subsamples of preceding random samples.

Multistage sampling can substantially reduce sampling costs, where the complete population list would need to be constructed (before other sampling methods could be applied). By eliminating the work involved in describing clusters that are not selected, multistage sampling can reduce the large costs associated with traditional cluster sampling. However, each sample may not be a full representative of the whole population.

Quota Sampling

In quota sampling, the population is first segmented into mutually exclusive subgroups, just as in stratified sampling. Then judgement is used to select the subjects or units from each segment based on a specified proportion. For example, an interviewer may be told to sample 200 females and 300 males between the age of 45 and 60.

It is this second step which makes the technique one of non-probability sampling. In quota sampling the selection of the sample is non-random. For example, interviewers might be tempted to interview those who look most helpful. The problem is that these samples may be biased because not everyone gets a chance of selection. This random element is its greatest weakness and quota versus probability has been a matter of controversy for several years.

Minimax Sampling

In imbalanced datasets, where the sampling ratio does not follow the population statistics, one can resample the dataset in a conservative manner called minimax sampling. The minimax sampling has its origin in Anderson minimax ratio whose value is proved to be 0.5: in a binary classification, the class-sample sizes should be chosen equally. This ratio can be proved to be minimax ratio only under the assumption of LDA classifier with Gaussian distributions. The notion of minimax sampling is recently developed for a general class of classification rules, called class-wise smart classifiers.

Accidental Sampling

Accidental sampling (sometimes known as grab, convenience or opportunity sampling) is a type of nonprobability sampling which involves the sample being drawn from that part of the population which is close to hand. That is, a population is selected because it is readily available and convenient. It may be through meeting the person or including a person in the sample when one meets them or chosen by finding them through technological means such as the internet or through phone. The researcher using such a sample cannot scientifically make generalizations about the total population from this sample because it would not be representative enough. For example, if the interviewer were to conduct such a survey at a shopping center early in the morning on a given day, the people that he/she could interview would be limited to those given there at that given time, which would not represent the views of other members of society in such an area, if the survey were to be conducted at different times of day and several times per week. This type of sampling is most useful for pilot testing. Several important considerations for researchers using convenience samples include:

1. Are there controls within the research design or experiment which can serve to lessen the impact of a non-random convenience sample, thereby ensuring the results will be more representative of the population?

2. Is there good reason to believe that a particular convenience sample would or should respond or behave differently than a random sample from the same population?

3. Is the question being asked by the research one that can adequately be answered using a convenience sample?

In social science research, snowball sampling is a similar technique, where existing study subjects are used to recruit more subjects into the sample. Some variants of snowball sampling, such as respondent driven sampling, allow calculation of selection probabilities and are probability sampling methods under certain conditions.

Line-intercept Sampling

Line-intercept sampling is a method of sampling elements in a region whereby an element is sampled if a chosen line segment, called a "transect", intersects the element.

Panel Sampling

Panel sampling is the method of first selecting a group of participants through a random sampling method and then asking that group for (potentially the same) information several times over a period of time. Therefore, each participant is interviewed at two or more time points; each period of data collection is called a "wave". The method was developed by sociologist Paul Lazarsfeld in 1938 as a means of studying political campaigns. This longitudinal sampling-method allows estimates of changes in the population, for example with regard to chronic illness to job stress to weekly food expenditures. Panel sampling can also be used to inform researchers about within-person health changes due to age or to help explain changes in continuous dependent variables such as spousal interaction. There have been several proposed methods of analyzing panel data, including MANOVA, growth curves, and structural equation modeling with lagged effects.

Snowball Sampling

Snowball sampling involves finding a small group of initial respondents and using them to recruit more respondents. It is particularly useful in cases where the population is hidden or difficult to enumerate.

Theoretical Sampling

Theoretical sampling occurs when samples are selected on the basis of the results of the data collected so far with a goal of developing a deeper understanding of the area or develop theory.

Replacement of Selected Units

Sampling schemes may be *without replacement* ('WOR'—no element can be selected more than once in the same sample) or *with replacement* ('WR'—an element may appear multiple times in the one sample). For example, if we catch fish, measure them, and immediately return them to the water before continuing with the sample, this is a WR design, because we might end up catching and measuring the same fish more than once. However, if we do not return the fish to the water (e.g., if we eat the fish), this becomes a WOR design.

Sample Size

Formulas, tables, and power function charts are well known approaches to determine sample size.

Steps for Using Sample Size Tables

1. Postulate the effect size of interest, α, and β.
2. Check sample size table
    - Select the table corresponding to the selected α
    - Locate the row corresponding to the desired power
    - Locate the column corresponding to the estimated effect size.
    - The intersection of the column and row is the minimum sample size required.

Sampling and Data Collection

Good data collection involves:

- Following the defined sampling process
- Keeping the data in time order
- Noting comments and other contextual events
- Recording non-responses

Applications of Sampling

Sampling enables the selection of right data points from within the larger data set to estimate the characteristics of the whole population. For example, there are about 600 million tweets produced every day. It is not necessary to look at all of them to determine the topics that are discussed during the day, nor is it necessary to look at all the tweets to determine the sentiment on each of the topics. A theoretical formulation for sampling Twitter data has been developed.

In manufacturing different types of sensory data such as acoustics, vibration, pressure, current, voltage and controller data are available at short time intervals. To predict down-time it may not be necessary to look at all the data but a sample may be sufficient.

Errors in Sample Surveys

Survey results are typically subject to some error. Total errors can be classified into sampling errors and non-sampling errors. The term "error" here includes systematic biases as well as random errors.

Sampling Errors and Biases

Sampling errors and biases are induced by the sample design. They include:

1. Selection bias: When the true selection probabilities differ from those assumed in calculating the results.
2. Random sampling error: Random variation in the results due to the elements in the sample being selected at random.

Non-sampling Error

Non-sampling errors are other errors which can impact the final survey estimates, caused by problems in data collection, processing, or sample design. They include:

1. Over-coverage: Inclusion of data from outside of the population.
2. Under-coverage: Sampling frame does not include elements in the population.
3. Measurement error: e.g. when respondents misunderstand a question, or find it difficult to answer.
4. Processing error: Mistakes in data coding.
5. Non-response: Failure to obtain complete data from all selected individuals.

After sampling, a review should be held of the exact process followed in sampling, rather than that intended, in order to study any effects that any divergences might have on subsequent analysis. A particular problem is that of *non-response.* Two major types of non-response exist: unit nonresponse (referring to lack of completion of any part of the survey) and item non-response (submission or participation in survey but failing to complete one or more components/questions of the survey). In survey sampling, many of the individuals identified as part of the sample may be unwilling to participate, not have the time to participate (opportunity cost), or survey administrators may not have been able to contact them. In this case, there is a risk of differences, between respondents and nonrespondents, leading to biased estimates of population parameters. This is often addressed by improving survey design, offering incentives, and conducting follow-up studies which make a repeated attempt to contact the unresponsive and to characterize their similarities and differences with the rest of the frame. The effects can also be mitigated by weighting the data when population benchmarks are available or by imputing data based on answers to other questions.

Nonresponse is particularly a problem in internet sampling. Reasons for this problem include improperly designed surveys, over-surveying (or survey fatigue), and the fact that potential participants hold multiple e-mail addresses, which they don't use anymore or don't check regularly.

Survey Weights

In many situations the sample fraction may be varied by stratum and data will have to be weighted to correctly represent the population. Thus for example, a simple random sample of individuals in the United Kingdom might include some in remote Scottish islands who would be inordinately expensive to sample. A cheaper method would be to use a stratified sample with urban and rural strata. The rural sample could be under-represented in the sample, but weighted up appropriately in the analysis to compensate.

More generally, data should usually be weighted if the sample design does not give each individual an equal chance of being selected. For instance, when households have equal selection probabilities but one person is interviewed from within each household, this gives people from large households a smaller chance of being interviewed. This can be accounted for using survey weights. Similarly, households with more than one telephone line have a greater chance of being selected in a random digit dialing sample, and weights can adjust for this.

Weights can also serve other purposes, such as helping to correct for non-response.

Methods of Producing Random Samples

- Random number table
- Mathematical algorithms for pseudo-random number generators
- Physical randomization devices such as coins, playing cards or sophisticated devices such as ERNIE

History

Random sampling by using lots is an old idea, mentioned several times in the Bible. In 1786 Pierre Simon Laplace estimated the population of France by using a sample, along with ratio estimator. He also computed probabilistic estimates of the error. These were not expressed as modern confidence intervals but as the sample size that would be needed to achieve a particular upper bound on the sampling error with probability 1000/1001. His estimates used Bayes' theorem with a uniform prior probability and assumed that his sample was random. Alexander Ivanovich Chuprov introduced sample surveys to Imperial Russia in the 1870s.

In the USA the 1936 *Literary Digest* prediction of a Republican win in the presidential election went badly awry, due to severe bias. More than two million people responded to the study with their names obtained through magazine subscription lists and telephone directories. It was not appreciated that these lists were heavily biased towards Republicans and the resulting sample, though very large, was deeply flawed.

Statistics is the Science of Data

Data are numerical values containing some information.

Statistical tools can be used on a data set to draw statistical inferences. These statistical inferences are in turn used for various purposes. For example, government uses such data for policy formulation for the welfare of the people, marketing companies use the data from consumer surveys to improve the company and to provide better services to the customer, etc. Such data is obtained through sample surveys. Sample surveys are conducted throughout the world by governmental as well as non-governmental agencies. For example, "National Sample Survey Organization (NSSO)" conducts surveys in India, "Statistics Canada" conducts surveys in Canada, agencies of United Nations like "World Health Organization (WHO), "Food and Agricultural Organization (FAO)" etc. conduct surveys in different countries.

Sampling theory provides the tools and techniques for data collection keeping in mind the objectives to be fulfilled and nature of population.

There are two ways of obtaining the information

1. Sample surveys
2. Complete enumeration or census

Sample surveys collect information on a fraction of total population whereas in census, the information is collected on the whole population. Some surveys, e.g. economic surveys, agricultural surveys etc. are conducted regularly. Some surveys are need based and are conducted when some need arises, e.g., consumer satisfaction surveys at a newly opened shopping mall to see the satisfaction level with the amenities provided in the mall.

Sampling Unit

An element or a group of elements on which observations can be taken is called a sampling unit. The objective of the survey helps in determining the definition of sampling unit.

For example, if the objective is to determine the total income of all the persons in the household, then the sampling unit is household. If the objective is to determine the income of any particular person in the household, then the sampling unit is the income of the particular person in the household. So the definition of sampling unit depends and varies as per the objective of the survey. Similarly, in another example, if the objective is to study the blood sugar level, then the sampling unit is the value of blood sugar level of a person. On the other hand, if the objective is to study the health conditions, then the sampling unit is the person on whom the readings on the blood sugar level, blood pressure and other factors will be obtained. These values will together classify the person as healthy or unhealthy.

Population

Collection of all the sampling units in a given region at a particular point of time or a particular period is called population. For example, if the medical facilities in a hospital are to be surveyed through the patients, then the total number of patients registered in the hospital during the time period of survey will be the population. Similarly, if the production of wheat in a district is to be studied, then all the fields cultivating wheat in that district will constitute the population. The total number of sampling units in the population is the population size, denoted generally by N. The population size can be finite or infinite (N is large).

Census

Census taker visits a Romani family living in a caravan, Netherlands 1925

A census is the procedure of systematically acquiring and recording information about the members of a given population. It is a regularly occurring and official count of a particular population. The term is used mostly in connection with national population and housing censuses; other common censuses include agriculture, business, and traffic censuses. The United Nations defines the essential features of population and housing censuses as "individual enumeration, universality within a defined territory, simultaneity and defined periodicity", and recommends that population censuses be taken at least every 10 years. United Nations recommendations also cover census topics to be collected, official definitions, classifications and other useful information to co-ordinate international practice.

The word is of Latin origin: during the Roman Republic, the census was a list that kept track of all adult males fit for military service. The modern census is essential to international comparisons of any kind of statistics, and censuses collect data on many attributes of a population, not just how many people there are but now census takes its place within a system of surveys where it typically began as the only national demographic data collection. Although population estimates remain an important function of a census, including exactly the geographic distribution of the population, statistics can be produced about combinations of attributes e.g. education by age and sex in different regions. Current administrative data systems allow for other approaches to enumeration with the same level of detail but raise concerns about privacy and the possibility of biasing estimates.

A census can be contrasted with sampling in which information is obtained only from a subset of a population, typically main population estimates are updated by such intercensal estimates. Modern census data are commonly used for research, business marketing, and planning, and as a baseline for designing sample surveys by providing a sampling frame such as an address register. Census counts are necessary to adjust samples to be representative of a population by weighting them as is common in opinion polling. Similarly, stratification requires knowledge of the relative sizes of different population strata which can be derived from census enumerations. In some countries, the census provides the official counts used to apportion the number of elected representatives to regions (sometimes controversially – e.g., *Utah v. Evans*). In many cases, a carefully chosen random sample can provide more accurate information than attempts to get a population census.

Sampling

A census is often construed as the opposite of a sample as its intent is to count everyone in a population rather than a fraction. However, population censuses rely on a sampling frame to count the population. This is the only way to be sure that everyone has been included as otherwise those not responding would not be followed up on and individuals could be missed. The fundamental premise of a census is that the population is not known and a new estimate is to be made by the analysis of primary data. The use of a sampling frame is counterintuitive as it suggests that the population size is already known. However, a census is also used to collect attribute data on the individuals in the nation. This process of sampling marks the difference between historical census, which was a house to house process or the product of an imperial decree, and the modern statistical project. The sampling frame used by census is almost always an address register. Thus it is not known if there is anyone resident or how many people there are in each household. Depending on the mode of enumeration, a form is sent to the householder, an enumerator calls, or administrative records for the dwelling are accessed. As a preliminary to the dispatch of forms, census workers will check any address problems on the ground. While it may seem straightforward to use the postal service file for this purpose, this can be out of date and some dwellings may contain a number of independent households. A particular problem is what are termed 'communal establishments' which category includes student residences, religious orders, homes for the elderly, people in prisons etc. As these are not easily enumerated by a single householder, they are often treated differently and visited by special teams of census workers to ensure they are classified appropriately.

Residence Definitions

Individuals are normally counted within households and information is typically collected about the household structure and the housing. For this reason international documents refer to censuses of population and housing. Normally the census response is made by a household, indicating details of individuals resident there. An important aspect of census enumerations is determining which individuals can be counted from which cannot be counted. Broadly, three definitions can be used: *de facto* residence; *de jure* residence; and, permanent residence. This is important to consider individuals who have multiple or temporary addresses. Every person should be identified uniquely as resident in one place but where they happen to be on Census Day, their *de facto* residence, may not be the best place to count them. Where an individual uses services may be more useful and this is at their usual, or *de jure*, residence. An individual may be represented at a permanent address, perhaps a family home for students or long term migrants. It is necessary to have a precise definition of residence to decide whether visitors to a country should be included in the population count. This is becoming more important as students travel abroad for education for a period of several years. Other groups causing problems of enumeration are new born babies, refugees, people away on holiday, people moving home around census day, and people without a fixed address. People having second homes because of working in another part of the country or retaining a holiday cottage are difficult to fix at a particular address sometimes causing double counting or houses being mistakenly identified as vacant. Another problem is where people use a different address at different times e.g. students living at their place of education in term time but returning to a family home during vacations or children whose parents have separated who effectively have two family homes. Census enumeration has always been based on finding people where they live as there is no systematic alternative - any list you could use to find people is derived from census activities in the first place. Recent UN guidelines provide recommendation on enumerating such complex households.

Enumeration Strategies

Historical censuses used crude enumeration assuming absolute accuracy. Modern approaches take into account the problems of overcount and undercount, and the coherence of census enumerations with other official sources of data. This reflects a realist approach to measurement, acknowledging that under any definition of residence there is a true value of the population but this can never be measured with complete accuracy. An important aspect of the census process is to evaluate the quality of the data.

Many countries use a post-enumeration survey to adjust the raw census counts. This works in a similar manner to capture-recapture estimation for animal populations. In census circles this method is called dual system enumeration (DSE). A sample of households are visited by interviewers who record the details of the household as at census day. These data are then matched to census records and the number of people missed can be estimated by considering the number missed in the census or survey but counted in the other. This way counts can be adjusted for non-response varying between different demographic groups. An explanation using a fishing analogy can be found in "Trout, Catfish and Roach..." which won an award from the Royal Statistical Society for excellence in official statistics in 2011.

Triple system enumeration has been proposed as an improvement as it would allow evaluation of the statistical dependence of pairs of sources. However, as the matching process is the most difficult aspect of census estimation this has never been implemented for a national enumeration. It would also be difficult to identify three different sources that were sufficiently different to make the triple system effort worthwhile. The DSE approach has another weakness in that it assumes there is no person counted twice (over count). In *de facto* residence definitions this would not be a problem but in *de jure* definitions individuals risk being recorded on more than one form leading to double counting. A particular problem here are students who often have a term time and family address.

Several countries have used a system which is known as short form/long form. This is a sampling strategy which randomly chooses a proportion of people to send a more detailed questionnaire to (the long form). Everyone receives the short form questions. Thereby more data are collected but not imposing a burden on the whole population. This also reduces the burden on the statistical office. Indeed, in the UK all residents were required to fill in the whole form but only a 10% sample were coded and analysed in detail, until 2001. New technology means that all data are now scanned and processed. Recently there has been controversy in Canada about the cessation of the long form with the head, Munir Sheikh resigning.

The use of alternative enumeration strategies is increasing but these are not so simple as many people assume and only occur in developed countries. The Netherlands has been most advanced in adopting a census using administrative data. This allows a simulated census to be conducted by linking several different administrative databases at an agreed time. Data can be matched and an overall enumeration established accounting for where the different sources are discrepant. A validation survey is still conducted in a similar way to the post enumeration survey employed in a traditional census. Other countries which have a population register use this as a basis for all the census statistics needed by users. This is most common amongst Nordic countries but requires a large number of different registers to be combined including population, housing, employment and education. These registers are then combined and brought up to the standard of a statistical register by comparing the data in different sources and ensuring the quality is sufficient for official statistics to be produced. A recent innovation is the French instigation of a rolling census programme with different regions enumerated each year such that the whole country is completely enumerated every 5 to 10 years. In Europe, in connection with the 2010 census round, a large number of countries adopted alternative census methodologies, often based on the combination of data from registers, surveys and other sources.

Technology

Censuses have evolved in their use of technology with the latest censuses, the 2010 round, using many new types of computing. In Brazil, handheld devices were used by enumerators to locate residences on the ground. In many countries, census returns could be made via the Internet as well as in paper form. DSE is facilitated by computer matching techniques which can be automated, such as propensity score matching. In the UK, all census formats are scanned and stored electronically before being destroyed, replacing the need for physical archives. The record linking to perform an administrative census would not be possible without large databases being stored on computer systems.

New technology is not without problems in its introduction. The US census had intended to use the handheld computers but cost escalated and this was abandoned, with the contract being sold to Brazil. Online response is a good idea but one of the functions of census is to make sure everyone is counted accurately. A system which allowed people to enter their address without verification would be open to abuse. Therefore, households have to be verified on the ground, typically by an enumerator visit or post out. Paper forms are still necessary for those without access to Internet connections. It is also possible that the hidden nature of an administrative census means that users are not engaged with the importance of contributing their data to official statistics.

Alternatively, population estimations may be carried out remotely with GIS and remote sensing technologies.

Census and Development

According to UNFPA, "The information generated by a population and housing census – numbers of people, their distribution, their living conditions and other key data – is critical for development." This is because this type of data is essential for policymakers so that they know where to invest. Unfortunately, many countries have outdated or inaccurate data about their populations and therefore, without accurate data are unable to address the needs of their population.

UNFPA stated that,

"The unique advantage of the census is that it represents the entire statistical universe, down to the smallest geographical units, of a country or region. Planners need this information for all kinds of development work, including: assessing demographic trends; analysing socio-economic conditions; designing evidence-based poverty-reduction strategies; monitoring and evaluating the effectiveness of policies; and tracking progress toward national and internationally agreed development goals."

In addition to making policymakers aware about population issues, it is also an important tool for identifying forms of social, demographic or economic exclusions, such as inequalities relating to race, ethics and religion as well as disadvantaged groups such as those with disabilities and the poor.

An accurate census can empower local communities by providing them with the necessary information to participate in local decision-making and ensuring they are represented.

Uses of Census Data

In the nineteenth century, the first censuses collected paper enumerations that had to be collated by hand so the statistical uses were very basic. The government owned the data and were able to publish statistics themselves on the state of the nation. Uses were to measure changes in the population and apportion representation. Population estimates could be compared to those of other countries.

By the beginning of the twentieth century, censuses were recording households and some indications of their employment. In some countries, census archives are released for public examination after many decades, allowing genealogists to track the ancestry of interested people. Archives provide a substantial historical record which may challenge established notions of tradition. It is also possible to understand the societal history through job titles and arrangements for the destitute and sick.

Census Data and Research

As governments assumed responsibility for schooling and welfare, large government research departments made extensive use of census data. Actuarial estimates could be made to project populations and plan for provision in local government and regions. It was also possible for central government to allocate funding on the basis of census data. Even into the mid twentieth century, census data was only directly accessible to large government departments. However, computers meant that tabulations could be used directly by university researchers, large businesses and local government offices. They could use the detail of the data to answer new questions and add to local and specialist knowledge.

Now, census data are published in a wide variety of formats to be accessible to business, all levels of governance, media, students and teachers, charities and any citizen who is interested; researchers in particular have an interest in the role of *Census Field Officers* (CFO) and their assistants. Data can be represented visually or analysed in complex statistical models, to show the difference between certain areas, or to understand the association between different personal characteristics. Census data offer a unique insight into small areas and small demographic groups which sample data would be unable to capture with precision.

Privacy

Although the census provides a useful way of obtaining statistical information about a population, such information can sometimes lead to abuses, political or otherwise, made possible by the linking of individuals' identities to anonymous census data. This consideration is particularly important when individuals' census responses are made available in microdata form, but even aggregate-level data can result in privacy breaches when dealing with small areas and/or rare subpopulations.

For instance, when reporting data from a large city, it might be appropriate to give the average income for black males aged between 50 and 60. However, doing this for a town that only has two black males in this age group would be a breach of privacy because either of those persons, knowing his own income and the reported average, could determine the other man's income.

Typically, census data are processed to obscure such individual information. Some agencies do this by intentionally introducing small statistical errors to prevent the identification of individuals in marginal populations; others swap variables for similar respondents. Whatever measures have been taken to reduce the privacy risk in census data, new technology in the form of better electronic analysis of data poses increasing challenges to the protection of sensitive individual information. This is known as statistical disclosure control.

Another possibility is to present survey results by means of statistical models in the form of a multivariate distribution mixture. The statistical information in the form of conditional distributions (histograms) can be derived interactively from the estimated mixture model without any further access to the original database. As the final product does not contain any protected microdata, the model based interactive software can be distributed without any confidentiality concerns.

Another method is simply to release no data at all, except very large scale data directly to the central government. Different release strategies between government have led to an international project (IPUMS) to co-ordinate access to microdata and corresponding metadata. Such projects also promote standardising metadata by projects such as SDMX so that best use can be made of the minimal data available.

Historical Censuses

Egypt

Censuses in Egypt first appears in the late Middle Kingdom and develops in the New Kingdom Pharaoh Amasis, according to Herodotus, require every Egyptian to declare annually to the nomarch, "whence he gained his living". Under the Ptolemies and the Romans several censuses were conducted in Egypt by governments officials

Ancient Greece

There are several accounts of ancient Greek city states carrying out censuses.

Ancient Israel

Censuses are mentioned in the Bible. God commands a per capita tax to be paid with the census in Exodus 30:11-16 for the upkeep of the Tabernacle. The Book of Numbers is named after the counting of the Israelite population (in Numbers 1-4) according to the house of the Fathers after the exodus from Egypt. A second census was taken while the Israelite were camped in the plains of Moab, in Numbers 26.

King David performed a census that produced disastrous results (in 2 Samuel 24 and 1 Chronicles 21). His son, King Solomon, had all of the foreigners in Israel counted in 2 Chronicles 2:17.

When the Romans took over Judea in AD 6, the legate Publius Sulpicius Quirinius organised a census for tax purposes. The Gospel of Luke links the birth of Jesus to this event. Luke 2.

China

One of the world's earliest preserved censuses was held in China in AD 2 during the Han Dynasty, and is still considered by scholars to be quite accurate. Another census was held in AD 144.

India

The oldest recorded census in India is thought to have occurred around 300 BC during the reign of the Emperor Chandragupta Maurya under the leadership of Kautilya or Chanakya and Ashoka.

Rome

The word "census" originated in ancient Rome from the Latin word *censere* ("to estimate"). The census played a crucial role in the administration of the Roman Empire, as it was used to determine taxes. With few interruptions, it was usually carried out every five years. It provided a register of citizens and their property from which their duties and privileges could be listed. It is said to have been instituted by the Roman king Servius Tullius in the 6th century BC, at which time the number of arms-bearing citizens was supposedly counted at around 80,000. The 6 BC "census of Quirinius" undertaken following the imposition of direct Roman rule in Judea was partially responsible for the development of the Zealot movement and several failed rebellions against Rome that ended in the Diaspora. The 15-year indiction cycle established by Diocletian in AD 297 was based on quindecennial censuses and formed the basis for dating in late antiquity and under the Byzantine Empire.

Rashidun and Umayyad Caliphates

In the Middle Ages, the Caliphate began conducting regular censuses soon after its formation, beginning with the one ordered by the second Rashidun caliph, Umar.

Medieval Europe

The Domesday Book was undertaken in AD 1086 by William I of England so that he could properly tax the land he had recently conquered in medieval Europe. In 1183, a census was taken of the crusader Kingdom of Jerusalem, to ascertain the number of men and amount of money that could possibly be raised against an invasion by Saladin, sultan of Egypt and Syria.

Inca Empire

In the 15th century, the Inca Empire had a unique way to record census information. The Incas did not have any written language but recorded information collected during censuses and other numeric information as well as non-numeric data on quipus, strings from llama or alpaca hair or cotton cords with numeric and other values encoded by knots in a base-10 positional system.

Spanish Empire

On May 25, 1577, King Philip II of Spain ordered by royal cédula the preparation of a general description of Spain's holdings in the Indies. Instructions and a questionnaire, issued in 1577 by the Office of the Cronista Mayor, were distributed to local officials in the Viceroyalties of New Spain and Peru to direct the gathering of information. The questionnaire, composed of fifty items, was designed to elicit basic information about the nature of the land and the life of its peoples. The replies, known as "relaciones geográficas," were written between 1579 and 1585 and were returned to the Cronista Mayor in Spain by the Council of the Indies.

World Population Estimates

TABLE 3.

PRESENT POPULATION OF THE EARTH AND THE CONTINENTS

(In Millions)

Continent	According to	
	International Statistical Institute 1929	League of Nations 1929
Asia	954	918
Europe	478	520
North America	162	161
Africa	140	146
South America	77	79
Australasia	9	9
Total	1,820	1,833

League of Nations and International Statistical Institute estimates of the world population in 1929

The earliest estimate of the world population was made by Giovanni Battista Riccioli in 1661; the next by Johann Peter Süssmilch in 1741, revised in 1762; the third by Karl Friedrich Wilhelm Dieterici in 1859.

In 1931 Walter Willcox published a table in his book, *International Migrations: Volume II Interpretations*, that estimated the 1929 world population to be roughly 1.8 billion.

Complete count of population is called census. The observations on all the sampling units in the population are collected in a census. For example, in India, the census is conducted at every tenth year in which observations on all the persons staying in India is collected.

Sample

One or more sampling units are selected from the population according to some specified procedure. A sample consists only of a portion of the population units.

In the context of sample surveys, a collection of units like households, people, cities, countries etc. is called a finite population.

A census is a 100% sample and it is a complete count of the population.

Representative Sample

All salient features of population are present in the sample.

It goes without saying that every sample is considered as a representative sample.

For example, if a population has 30% males and 70% females, then we also expect the sample to have nearly 30% males and 70% females.

In another example, if we take out a handful of wheat from a 100 kg. bag of wheat, we expect the same quality of wheat in hand as inside the bag. Similarly, it is expected that a drop of blood will give the same information as all the blood in the body.

Sampling Frame

In statistics, a sampling frame is the source material or device from which a sample is drawn. It is a list of all those within a population who can be sampled, and may include individuals, households or institutions.

Importance of the sampling frame is stressed by Jessen and Salant and Dillman.

In many practical situations the frame is a matter of choice to the survey planner, and sometimes a critical one. [...] Some very worthwhile investigations are not undertaken at all because of the lack of an apparent frame; others, because of faulty frames, have ended in a disaster or in cloud of doubt.

—*Raymond James Jessen*

The white pages, a collection of telephone directories in the United States, was commonly used as a sampling frame for opinion polls.

Obtaining and Organizing a Sampling Frame

In the most straightforward cases, such as when dealing with a batch of material from a production run, or using a census, it is possible to identify and measure every single item in the population and to include any one of them in our sample; this is known as *direct element sampling*. However, in many other cases this is not possible; either because it is cost-prohibitive (reaching every citizen of a country) or impossible (reaching all humans alive).

Having established the frame, there are a number of ways for organizing it to improve efficiency and effectiveness. It's at this stage that the researcher should decide whether the sample is in fact to be the whole population and would therefore be a census.

This list should also facilitate access to the selected sampling units. A frame may also provide additional 'auxiliary information' about its elements; when this information is related to variables or groups of interest, it may be used to improve survey design. While not necessary for simple sampling, a sampling frame used for more advanced sample techniques, such as stratified sampling, may contain additional information (such as demographic information). For instance, an electoral register might include name and sex; this information can be used to ensure that a sample taken from that frame covers all demographic categories of interest. (Sometimes the auxiliary information is less explicit; for instance, a telephone number may provide some information about location).

Sampling Frame Qualities

An ideal sampling frame will have the following qualities:

- all units have a logical, numerical identifier
- all units can be found – their contact information, map location or other relevant information is present
- the frame is organized in a logical, systematic fashion
- the frame has additional information about the units that allow the use of more advanced sampling frames
- every element of the population of interest is present in the frame
- every element of the population is present *only once* in the frame
- no elements from outside the population of interest are present in the frame
- the data is 'up-to-date'

Types of Sampling Frames

The most straightforward type of frame is a list of elements of the population (preferably the entire population) with appropriate contact information. For example, in an opinion poll, possible sampling frames include an electoral register or a telephone directory. Other sampling frames can include employment records, school class lists, patient files in a hospital, organizations listed in a thematic database, and so on. On a more practical levels, sampling frames have the form of computer files.

Not all frames explicitly list population elements; some list only 'clusters'. For example, a street map can be used as a frame for a door-to-door survey; although it doesn't show individual houses, we can select streets from the map and then select houses on those streets. This offers some advantages: such a frame would include people who have recently moved and are not yet on the list frames discussed above, and it may be easier to use because it doesn't require storing data for every unit in the population, only for a smaller number of clusters.

Sampling Frames Problems

The sampling frame must be representative of the population and this is a question outside the scope of statistical theory demanding the judgment of experts in the particular subject matter being studied. All the above frames omit some people who will vote at the next election and contain some people who will not; some frames will contain multiple records for the same person. People not in the frame have no prospect of being sampled.

Because a cluster-based frame contains less information about the population, it may place constraints on the sample design, possibly requiring the use of less efficient sampling methods and/or making it harder to interpret the resulting data.

Statistical theory tells us about the uncertainties in extrapolating from a sample to the frame. It should be expected that sample frames, will always contain some mistakes. In some cases, this may lead to sampling bias. Such bias should be minimized, and identified, although avoiding it completely in a real world is nearly impossible. One should also not assume that sources which claim to be unbiased and representative are such.

In defining the frame, practical, economic, ethical, and technical issues need to be addressed. The need to obtain timely results may prevent extending the frame far into the future. The difficulties can be extreme when the population and frame are disjoint. This is a particular problem in forecasting where inferences about the future are made from historical data. In fact, in 1703, when Jacob Bernoulli proposed to Gottfried Leibniz the possibility of using historical mortality data to predict the probability of early death of a living man, Gottfried Leibniz recognized the problem in replying:

> Nature has established patterns originating in the return of events but only for the most part. New illnesses flood the human race, so that no matter how many experiments you have done on corpses, you have not thereby imposed a limit on the nature of events so that in the future they could not vary.
>
> —*Gottfried Leibniz*

Leslie Kish posited four basic problems of sampling frames:

1. Missing elements: Some members of the population are not included in the frame.
2. Foreign elements: The non-members of the population are included in the frame.
3. Duplicate entries: A member of the population is surveyed more than once.
4. Groups or clusters: The frame lists clusters instead of individuals.

Problems like those listed can be identified by the use of pre-survey tests and pilot studies.

List of all the units of the population to be surveyed constitutes the sampling frame. All the sampling units in the sampling frame have identification particulars. For example, all the students in a particular university listed along with their roll numbers constitute the sampling frame. Similarly, the list of households with the name of head of family or house address constitutes the sampling frame. In another example, the residents of a city area may be listed in more than one frame - as per automobile registration as well as the listing in the telephone directory.

Ways to Ensure Representativeness

There are two possible ways to ensure that the selected sample is representative.

1. Random sample or probability sample:

The selection of units in the sample from a population is governed by the laws of chance or probability. The probability of selection of a unit can be equal as well as unequal.

2. Non-random sample or purposive sample:

The selection of units in the sample from population is not governed by the probability laws.

For example, the units are selected on the basis of personal judgment of the surveyor. The persons volunteering to take some medical test or to drink a new type of coffee also constitute the sample on non- random laws.

Another type of sampling is Quota Sampling. The survey in this case is continued until a predetermined number of units with the characteristic under study are picked up.

For example, in order to conduct an experiment for rare type of disease, the survey is continued till the required number of patients with disease are collected.

Advantages of Sampling Over Complete Enumeration

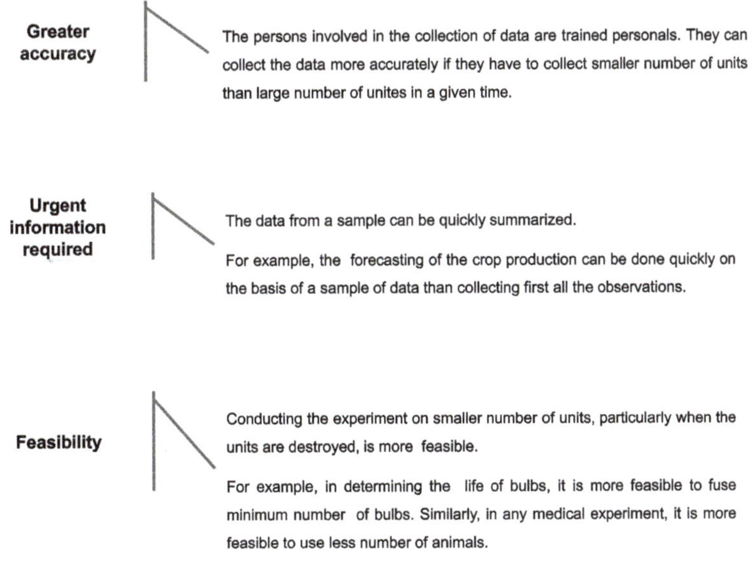

Greater accuracy
The persons involved in the collection of data are trained personals. They can collect the data more accurately if they have to collect smaller number of units than large number of unites in a given time.
Urgent information required
The data from a sample can be quickly summarized.
For example, the forecasting of the crop production can be done quickly on the basis of a sample of data than collecting first all the observations.
Feasibility
Conducting the experiment on smaller number of units, particularly when the units are destroyed, is more feasible.
For example, in determining the life of bulbs, it is more feasible to fuse minimum number of bulbs. Similarly, in any medical experiment, it is more feasible to use less number of animals.

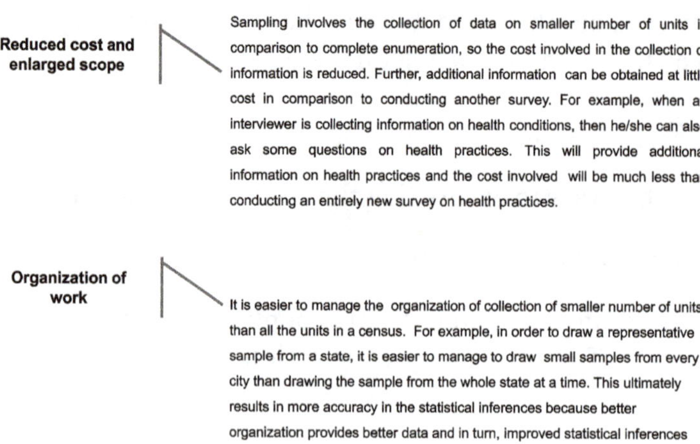

Reduced cost and enlarged scope
Sampling involves the collection of data on smaller number of units in comparison to complete enumeration, so the cost involved in the collection of information is reduced. Further, additional information can be obtained at little cost in comparison to conducting another survey. For example, when an interviewer is collecting information on health conditions, then he/she can also ask some questions on health practices. This will provide additional information on health practices and the cost involved will be much less than conducting an entirely new survey on health practices.
Organization of work
It is easier to manage the organization of collection of smaller number of units than all the units in a census. For example, in order to draw a representative sample from a state, it is easier to manage to draw small samples from every city than drawing the sample from the whole state at a time. This ultimately results in more accuracy in the statistical inferences because better organization provides better data and in turn, improved statistical inferences are obtained.

Survey Methodology

A field of applied statistics of human research surveys, survey methodology studies the sampling of individual units from a population and the associated survey data collection techniques, such as questionnaire construction and methods for improving the number and accuracy of responses to surveys. Survey methodology includes instruments or procedures that ask one or more questions that may, or may not, be answered.

Statistical surveys are undertaken with a view towards making statistical inferences about the population being studied, and this depends strongly on the survey questions used. Polls about public opinion, public health surveys, market research surveys, government surveys and censuses are all examples of quantitative research that use contemporary survey methodology to answer questions about a population. Although censuses do not include a "sample," they do include other aspects of survey methodology, like questionnaires, interviewers, and nonresponse follow-up techniques. Surveys provide important information for all kinds of public information and research fields, e.g., marketing research, psychology, health professionals and sociology.

A single survey is made of at least a sample (or full population in the case of a census), a method of data collection (e.g., a questionnaire) and individual questions or items that become data that can be analyzed statistically. A single survey may focus on different types of topics such as preferences (e.g., for a presidential candidate), opinions (e.g., should abortion be legal?), behavior (smoking and alcohol use), or factual information (e.g., income), depending on its purpose. Since survey research is almost always based on a sample of the population, the success of the research is dependent on the representativeness of the sample with respect to a target population of interest to the researcher.

That target population can range from the general population of a given country to specific groups of people within that country, to a membership list of a professional organization, or list of students enrolled in a school system. The persons replying to a survey are called respondents, and depending on the questions asked their answers may represent themselves as individuals, their households, employers, or other organization they represent.

Survey methodology as a scientific field seeks to identify principles about the sample design, data collection instruments, statistical adjustment of data, and data processing, and final data analysis that can create systematic and random survey errors. Survey errors are sometimes analyzed in connection with survey cost. Cost constraints are sometimes framed as improving quality within cost constraints, or alternatively, reducing costs for a fixed level of quality. Survey methodology is both a scientific field and a profession, meaning that some professionals in the field focus on survey errors empirically and others design surveys to reduce them. For survey designers, the task involves making a large set of decisions about thousands of individual features of a survey in order to improve it.

The most important methodological challenges of a survey methodologist include making decisions on how to:

- Identify and select potential sample members.
- Contact sampled individuals and collect data from those who are hard to reach (or reluctant to respond)
- Evaluate and test questions.
- Select the mode for posing questions and collecting responses.
- Train and supervise interviewers (if they are involved).
- Check data files for accuracy and internal consistency.
- Adjust survey estimates to correct for identified errors.

Selecting Samples

The sample is chosen from the sampling frame, which consists of a list of all members of the population of interest. The goal of a survey is not to describe the sample, but the larger population. This generalizing ability is dependent on the representativeness of the sample, as stated above. Each member of the population is termed an element. There are frequent difficulties one encounters while choosing a representative sample. One common error that results is selection bias. Selection bias results when the procedures used to select a sample result in over representation or under representation of some significant aspect of the population. For instance, if the population of interest consists of 75% females, and 25% males, and the sample consists of 40% females and 60% males, females are under represented while males are overrepresented. In order to minimize selection biases, stratified random sampling is often used. This is when the population is divided into sub-populations called strata, and random samples are drawn from each of the strata, or elements are drawn for the sample on a proportional basis.

Modes of Data Collection

There are several ways of administering a survey. The choice between administration modes is influenced by several factors, including

1. costs,
2. coverage of the target population,
3. flexibility of asking questions,
4. respondents' willingness to participate and
5. response accuracy.

Different methods create mode effects that change how respondents answer, and different methods have different advantages. The most common modes of administration can be summarized as:

- Telephone
- Mail (post)
- Online surveys
- Personal in-home surveys
- Personal mall or street intercept survey
- Hybrids of the above.

Research Designs

There are several different designs, or overall structures, that can be used in survey research. The three general types are cross-sectional, successive independent samples, and longitudinal studies.

Cross-sectional Studies

In cross-sectional studies, a sample (or samples) is drawn from the relevant population and studied once. A cross-sectional study describes characteristics of that population at one time, but cannot give any insight as to the causes of population characteristics because it is a predictive, correlational design.

Successive Independent Samples Studies

A successive independent samples design draws multiple random samples from a population at one or more times. This design can study changes within a population, but not changes within individuals because the same individuals are not surveyed more than once. Such studies cannot, therefore, identify the causes of change over time necessarily. For successive independent samples designs to be effective, the samples must be drawn from the same population, and must be equally representative of it. If the samples are not comparable, the changes between samples may be due to demographic characteristics rather than time. In addition, the questions must be asked in the same way so that responses can be compared directly.

Longitudinal Studies

Longitudinal studies take measure of the same random sample at multiple time points. Unlike with a successive independent samples design, this design measures the differences in individual participants' responses over time. This means that a researcher can potentially assess the reasons for response changes by assessing the differences in respondents' experiences. Longitudinal studies are the easiest way to assess the effect of a naturally occurring event, such as divorce that cannot be tested experimentally. However, longitudinal studies are both expensive and difficult to do. It's harder to find a sample that will commit to a months- or years-long study than a 15-minute interview, and participants frequently leave the study before the final assessment. This attrition of participants is not random, so samples can become less representative with successive assessments. To account for this, a researcher can compare the respondents who left the survey to those that did not, to see if they are statistically different populations. Respondents may also try to be self-consistent in spite of changes to survey answers.

Questionnaires

Questionnaires are the most commonly used tool in survey research. However, the results of a particular survey are worthless if the questionnaire is written inadequately. Questionnaires should produce valid and reliable demographic variable measures and should yield valid and reliable individual disparities that self-report scales generate.

Questionnaires as Tools

A variable category that is often measured in survey research are demographic variables, which are used to depict the characteristics of the people surveyed in the sample. Demographic variables include such measures as ethnicity, socioeconomic status, race, and age. Surveys often assess the preferences and attitudes of individuals, and many employ self-report scales to measure people's opinions and judgements about different items presented on a scale. Self-report scales are also used to examine the disparities among people on scale items. These self-report scales, which are usually presented in questionnaire form, are one of the most used instruments in psychology, and thus it is important that the measures be constructed carefully, while also being reliable and valid.

Reliability and Validity of Self-report Measures

Reliable measures of self-report are defined by their consistency. Thus, a reliable self-report measure produces consistent results every time it is executed. A test's reliability can be measured a few ways. First, one can calculate a test-retest reliability. A test-retest reliability entails conducting the same questionnaire to a large sample at two different times. For the questionnaire to be considered reliable, people in the sample do not have to score identically on each test, but rather their position in the score distribution should be similar for both the test and the retest. Self-report measures will generally be more reliable when they have many items measuring a construct. Furthermore, measurements will be more reliable when the factor being measured has greater variability among the individuals in the sample that are being tested. Finally, there will be greater reliability when instructions for the completion of the questionnaire are clear and when there are limited distractions in the testing environment. Contrastingly, a questionnaire is valid if what it measures is what it had originally planned to measure. Construct validity of a measure is the degree to which it measures the theoretical construct that it was originally supposed to measure.

Composing a Questionnaire

Six steps can be employed to construct a questionnaire that will produce reliable and valid results. First, one must decide what kind of information should be collected. Second, one must decide how to conduct the questionnaire. Thirdly, one must construct a first draft of the questionnaire. Fourth, the questionnaire should be revised. Next, the questionnaire should be pretested. Finally, the questionnaire should be edited and the procedures for its use should be specified.

Guidelines for the Effective Wording of Questions

The way that a question is phrased can have a large impact on how a research participant will answer the question. Thus, survey researchers must be conscious of their wording when writing survey questions. It is important for researchers to keep in mind that different individuals, cultures, and subcultures can interpret certain words and phrases differently from one another. There are two different types of questions that survey researchers use when writing a questionnaire: free response questions and closed questions. Free response questions are open-ended, whereas closed questions are usually multiple choice. Free response questions are beneficial because they allow the responder greater flexibility, but they are also very difficult to record and score, requiring extensive coding. Contrastingly, closed questions can be scored and coded much easier, but they diminish expressivity and spontaneity of the responder. In general, the vocabulary of the questions should be very simple and direct, and most should be less than twenty words. Each question should be edited for "readability" and should avoid leading or loaded questions. Finally, if multiple items are being used to measure one construct, the wording of some of the items should be worded in the opposite direction to evade response bias.

A respondent's answer to an open-ended question can be coded into a response scale afterwards, or analysed using more qualitative methods.

Order of Questions

Survey researchers should carefully construct the order of questions in a questionnaire. For questionnaires that are self-administered, the most interesting questions should be at the beginning of the questionnaire to catch the respondent's attention, while demographic questions should be near the end. Contrastingly, if a survey is being administered over the telephone or in person, demographic questions should be administered at the beginning of the interview to boost the respondent's confidence. Another reason to be mindful of question order may cause a survey response effect in which one question may affect how people respond to subsequent questions as a result of priming.

Nonresponse Reduction

The following ways have been recommended for reducing nonresponse in telephone and face-to-face surveys:

- Advance letter. A short letter is sent in advance to inform the sampled respondents about the upcoming survey. The style of the letter should be personalized but not overdone. First, it announces that a phone call will be made/ or an interviewer wants to make an appointment to do the survey face-to-face. Second, the research topic will be described. Last, it allows both an expression of the surveyor's appreciation of cooperation and an opening to ask questions on the survey.
- Training. The interviewers are thoroughly trained in how to ask respondents questions, how to work with computers and making schedules for callbacks to respondents who were not reached.
- Short introduction. The interviewer should always start with a short introduction about him or herself. She/he should give her name, the institute she is working for, the length of the interview and goal of the interview. Also it can be useful to make clear that you are not selling anything: this has been shown to lead to a slightly higher responding rate.

- Respondent-friendly survey questionnaire. The questions asked must be clear, non-offensive and easy to respond to for the subjects under study.

Brevity is also often cited as increasing response rate. A 1996 literature review found mixed evidence to support this claim for both written and verbal surveys, concluding that other factors may often be more important. A 2010 study looking at 100,000 online surveys found response rate dropped by about 3% at 10 questions and about 6% at 20 questions, with drop-off slowing (for example, only 10% reduction at 40 questions). Other studies showed that quality of response degraded toward the end of long surveys.

Interviewer Effects

Survey methodologists have devoted much effort to determining the extent to which interviewee responses are affected by physical characteristics of the interviewer. Main interviewer traits that have been demonstrated to influence survey responses are race, gender, and relative body weight (BMI). These interviewer effects are particularly operant when questions are related to the interviewer trait. Hence, race of interviewer has been shown to affect responses to measures regarding racial attitudes, interviewer sex responses to questions involving gender issues, and interviewer BMI answers to eating and dieting-related questions. While interviewer effects have been investigated mainly for face-to-face surveys, they have also been shown to exist for interview modes with no visual contact, such as telephone surveys and in video-enhanced web surveys. The explanation typically provided for interviewer effects is social desirability bias: survey participants may attempt to project a positive self-image in an effort to conform to the norms they attribute to the interviewer asking questions. Interviewer effects are one example survey response effects.

Type of Surveys

1. Demographic surveys

These surveys are conducted to collect the demographic data, e.g., household surveys, family size, number of males in families, etc.

Such surveys are useful in the policy formulation for any city, state or country for the welfare of the people.

2. Educational surveys

These surveys are conducted to collect the educational data, e.g., how many children go to school, how many persons are graduate, etc.

Such surveys are conducted to examine the educational programs in schools and colleges. Generally, schools are selected first and then the students from each school constitue the sample.

3. Economic surveys

These surveys are conducted to collect the economic data, e.g., data related to export and import of goods, industrial production, consumer expenditure etc. Such data is helpful in constructing the indices indicating the growth in a particular sector of economy or even the overall economic growth of the country.

4. Employment surveys

These surveys are conducted to collect the employment related data, e.g., employment rate, labour conditions, wages, etc. in a city, state or country. Such data helps in constructing various indices to know the employment conditions among the people.

5. Health and nutrition surveys

These surveys are conducted to collect the data related to health and nutrition issues, e.g., number of visits to doctors, food given to children, nutritional value etc. Such surveys are conducted in cities, states as well as countries by national and international organizations like UNICEF, WHO etc.

6. Agricultural surveys

These surveys are conducted to collect the agriculture related data to estimate, e.g., the acreage and production of crops, livestock numbers, use of fertilizers, use of pesticides and other related topics. The government bases its planning related to the food issues for the people based on such surveys.

7. Marketing surveys

These surveys are conducted to collect the data related to marketing. They are conducted by major companies, manufacturers or those who provide services to consumer etc. Such data is used for knowing the satisfaction and opinion of consumers as well as in developing the sales, purchase and promotional activities etc.

8. Election surveys

These surveys are conducted to study the outcome of an election or a poll. For example, such polls are conducted in democratic countries to have the opinions of people about any candidate who is contesting the election.

9. Public polls and surveys

These surveys are conducted to collect the public opinion on any particular issue. Such surveys are generally conducted by news media and agencies which conduct polls and surveys on current topics of interest to public.

10. Campus surveys

These surveys are conducted on the students of any educational institution to study about the educational programs, living facilities, dining facilities, sports activities, etc.

Principal Steps in a Sample Survey

The broad steps to conduct any sample survey are as follows:

Objective of the Survey

The objective of the survey has to be clearly defined and well understood by the person planning to conduct it. It is expected from the statistician to be well versed with the issues to be addressed in consultation with the person who wants to get the survey conducted. In complex surveys, sometimes the objective is forgotten and data is collected on those issues which are far away from the objectives.

Population to be Sampled

Based on the objectives of the survey, decide the population from which the information can be obtained. For example, population of farmers is to be sampled for an agricultural survey whereas the population of patients has to be sampled for determining the medical facilities in a hospital.

Data to be Collected

It is important to decide that which data is relevant for fulfilling the objectives of the survey and to note that no essential data is omitted. Sometimes, too many questions are asked and some of their outcomes are never utilized. This lowers the quality of the responses and in turn results in lower efficiency in statistical inferences.

Degree of Precision Required

The results of any sample survey are always subjected to some uncertainty. Such uncertainty can be reduced by taking larger samples or using superior instruments. This involves more cost and more time. So, it is very important to decide about the required degree of precision in data. This needs to be conveyed to the surveyor also.

Method of Measurement

The choice of measuring instrument and method to measure the data from the population needs to be specified clearly. For example, the data has to be collected through interview, questionnaire, personal visit, combination of any of these approaches, etc. The forms in which the data is to be recorded so that the data can be transferred to mechanical equipment for easily creating the data summary etc. is also needed to be prepared accordingly.

The Frame

The sampling frame has to be clearly specified. The population is divided into sampling units such that the units cover the whole population and every sampling unit is tagged with identification. The list of all sampling units is called the frame. The frame must cover the whole population and the units must not overlap each other in the sense that every element in the population must belong to one and only one unit. For example, the sampling unit can be an individual member in the family or the whole family.

Selection of Sample

The size of the sample needs to be specified for the given sampling plan. This helps in determining and comparing the relative cost and time of different sampling plans. The method and plan adopted for drawing a representative sample should also be detailed.

The Pre-test

It is advised to try the questionnaire and field methods on a small scale. This may reveal some troubles and problems beforehand which the surveyor may face in field in large scale surveys.

Organization of the Field Work

How to conduct the survey, how to handle business administrative issues, providing proper training to surveyors, procedures, plans for handling the nonresponse and missing observations etc. are some of the issues which need to be addressed for organizing the survey work in the fields. The procedure for early checking of the quality of return should be prescribed. It should be clarified how to handle the situation when the respondent is not available.

Analysis of Data

It is to be noted that based on the objectives of the data, the suitable statistical tool is decided which can answer the relevant questions. In order to use the statistical tool, a valid data set is required and this dictates the choice of responses to be obtained for the questions in the questionnaire, e.g., the data has to be qualitative, quantitative, nominal, ordinal etc. After getting the completed questionnaire back, it needs to be edited to amend the recording errors and delete the erroneous data. The tabulating procedures, methods of estimation and tolerable amount of error in the estimation needs to be decided before the start of survey. Different methods of estimation may be available to get the answer of the same query from the same data set. So the data needs to be collected which is compatible with the chosen estimation procedure.

Information Gained for Future Surveys

The completed surveys work as guide for improved sample surveys in future. Beside this they also supply various types of prior information required to use various statistical tools, e.g., mean, variance, nature of variability, cost involved etc. Any completed sample survey acts as a potential guide for the surveys to be conducted in the future. It is generally seen that the things always do not go in the same way in any complex survey as planned earlier. Such precautions and alerts help in avoiding the mistakes in the execution of future surveys.

Variability Control in Sample Surveys

The variability control is an important issue in any statistical analysis. A general objective is to draw statistical inferences with minimum variability. There are various types of sampling schemes which are adopted in different conditions. These schemes help in controlling the variability at different stages. Such sampling schemes can be classified in the following way.

Before Selection of Sampling Units

- Stratified sampling
- Cluster sampling
- Two stage sampling
- Double sampling etc.

At the Time of Selection of Sampling Units

- Systematic sampling
- Varying probability sampling

After the Selection of Sampling Units

- Ratio method of estimation
- Regression method of estimation

Methods of Data Collection

Physical Observations and Measurements:

The surveyor contacts the respondent personally through meeting. He observes the sampling unit and records the data. The surveyor can always use his prior experience to collect the data in a better way. For example, a young man telling his age as 60 years can easily be observed and corrected by the surveyor.

Personal Interview:

The surveyor is supplied with a well prepared questionnaire. The surveyor goes to the respondents and asks the same questions mentioned in the questionnaire. The data in the questionnaire is then filled up accordingly based on the responses from the respondents.

Mail Enquiry:

The well prepared questionnaire is sent to the respondents through postal mail, e-mail, etc. The respondents are requested to fill up the questionnaires and send it back. In case of postal mail, many times the questionnaires are accompanied by a self addressed envelope with postage stamps to avoid any non-response due to the cost of postage.

Web Based Enquiry:

The survey is conducted online through internet based web pages. There are various websites which provide such facility. The questionnaires are to be in their formats and the link is sent to the respondents through email. By clicking on the link, the respondent is brought to the concerned website and the answers are to be given online. These answers are recorded and responses as well as their statistics is sent to the surveyor. The respondents should have internet connection to support the data collection with this procedure.

Registration:

The respondent is required to register the data at some designated place. For example, the number of births and deaths along with the details provided by the family members are recorded at city municipal office which are provided by the family members.

Transcription From Records:

The sample of data is collected from the already recorded information. For example, the details of the number of persons in different families or number of births/ deaths in a city can be obtained from the city municipal office directly. The methods in (1) to (5) provide primary data which means collecting the data directly from the source. The method in (6) provides the secondary data which means getting the data from the primary sources.

Statistical Unit

A unit in a statistical analysis refers to one member of a set of entities being studied. It is the material source for the mathematical abstraction of a "random variable". Common examples of a unit would be a single person, animal, plant, manufactured item, or country that belongs to a larger collection of such entities being studied.

Units are often referred to as being either experimental units, sampling units or, more generally, units of observation:

- An "experimental unit" is typically thought of as one member of a set of objects that are initially equivalent, with each object then subjected to one of several experimental treatments.
- A "sampling unit" is typically thought of as an object that has been sampled from a statistical population. This term is commonly used in opinion polling and survey sampling.

In most statistical studies, the goal is to generalize from the observed units to a larger set consisting of all comparable units that exist but are not directly observed. For example, if we randomly sample 100 people and ask them which candidate they intend to vote for in an election, our main interest is in the voting behavior of all eligible voters, not exclusively on the 100 observed units.

In some cases, the observed units may not form a sample from any meaningful population, but rather constitute a convenience sample, or may represent the entire population of interest. In this situation, we may study the units descriptively, or we may study their dynamics over time. But it typically does not make sense to talk about generalizing to a larger population of such units. Studies involving countries or business firms are often of this type. Clinical trials also typically use convenience samples, however the aim is often to make inferences about the efficacy of treatments in other patients, and given the inclusion and exclusion criteria for some clinical trials, the sample may not be representative of the majority of patients with the condition or disease.

In simple data sets, the units are in one-to-one correspondence with the data values. In more complex data sets, multiple measurements are made for each unit. For example, if blood pressure measurements are made daily for a week on each subject in a study, there would be seven data values for each statistical unit. Multiple measurements taken on an individual are not independent (they will be more alike compared to measurements taken on different individuals). Ignoring these dependencies during the analysis can lead to an inflated sample size or pseudoreplication.

While a *unit* is often the lowest level at which observations are made, in some cases, a *unit* can be further decomposed as a statistical assembly.

Many statistical analyses use quantitative data that have units of measurement. This is a distinct and non-overlapping use of the term "unit."

References

- Flores-Macias, F.; Lawson, C. (2008). "Effects of interviewer gender on survey responses: Findings from a household survey in Mexico". International Journal of Public Opinion Research. 20 (1): 100–110. doi:10.1093/ijpor/edn007
- Kane, E.W.; MacAulay, L.J. (1993). "Interviewer gender and gender attitudes". Public Opinion Quarterly. 57 (1): 1–28. doi:10.1086/269352

- Roger Sapsford; Victor Jupp (29 March 2006). Data collection and analysis. SAGE. pp. 28–. ISBN 978-0-7619-4363-1. Retrieved 2 January 2011
- Eisinga, R.; Te Grotenhuis, M.; Larsen, J.K.; Pelzer, B.; Van Strien, T. (2011). "BMI of interviewer effects". International Journal of Public Opinion Research. 23 (4): 530–543. doi:10.1093/ijpor/edr026
- Hill, M.E (2002). "Race of the interviewer and perception of skin color: Evidence from the multi-city study of urban inequality". American Sociological Review. 67 (1): 99–108. JSTOR 3088935. doi:10.2307/3088935
- Carl-Erik Särndal; Bengt Swensson; Jan Wretman (2003). Model assisted survey sampling. Springer. pp. 9–12. ISBN 978-0-387-40620-6. Retrieved 2 January 2011
- Scott, A.J.; Wild, C.J. (1986). "Fitting logistic models under case-control or choice-based sampling". Journal of the Royal Statistical Society, Series B. 48: 170–182. JSTOR 2345712
- Shahrokh Esfahani, Mohammad; Dougherty, Edward (2014). "Effect of separate sampling on classification accuracy". Bioinformatics. 30 (2): 242–250. PMID 24257187. doi:10.1093/bioinformatics/btt662
- Groves, R.M.; Fowler, F. J.; Couper, M.P.; Lepkowski, J.M.; Singer, E.; Tourangeau, R. (2009). Survey Methodology. New Jersey: John Wiley & Sons. ISBN 978-1-118-21134-2

2

Simple Random Sampling: An Overview

Simple random sampling is the collection of data taken from a large population. Programming and discrete uniform distribution are some of the methods by which the processes related to simple random sampling can be implemented. The chapter strategically encompasses and incorporates the major components and key concepts of simple random sampling, providing a complete understanding.

Simple Random Sampling

Simple random sampling (SRS) is a method of selection of a sample comprising of n number of sampling units from the population having N number of units such that every sampling unit has an equal chance of being chosen.

The Samples Can be Drawn in Two Possible Ways

- The sampling units are chosen without replacement in the sense that the units once chosen are not placed back in the population.
- The sampling units are chosen with replacement in the sense that the chosen units are placed back in the population.

Based on these two concepts, there are two approaches for SRS:

Simple Random Sampling without Replacement (SRSWOR)

SRSWOR is a method of selection of n units out of the N units one by one such that at any stage of selection, anyone of the remaining units have same chance of being selected, i.e., 1/N.

Simple Random Sampling with Replacement (SRSWR)

SRSWR is a method of selection of n units out of the N units one by one such that at each stage of selection each unit has equal chance of being selected, i.e., 1/N.

In statistics, a simple random sample is a subset of individuals (a sample) chosen from a larger set (a population). Each individual is chosen randomly and entirely by chance, such that each individual has the same probability of being chosen at any stage during the sampling process, and each subset of *k* individuals has the same probability of being chosen for the sample as any other subset of k individuals. This process and technique is known as simple random sampling, and should not be confused with systematic random sampling. A simple random sample is an unbiased surveying technique.

Simple random sampling is a basic type of sampling, since it can be a component of other more complex sampling methods. The principle of simple random sampling is that every object has the same probability of being chosen. For example, suppose N college students want to get a ticket for a basketball game, but there are only $X < N$ tickets for them, so they decide to have a fair way to see who gets to go. Then, everybody is given a number in the range from 0 to N-1, and random numbers are generated, either electronically or from a table of random numbers. Numbers outside the range from 0 to N-1 are ignored, as are any numbers previously selected. The first X numbers would identify the lucky ticket winners.

In small populations and often in large ones, such sampling is typically done "without replacement", i.e., one deliberately avoids choosing any member of the population more than once. Although simple random sampling can be conducted with replacement instead, this is less common and would normally be described more fully as simple random sampling with replacement. Sampling done without replacement is no longer independent, but still satisfies exchangeability, hence many results still hold. Further, for a small sample from a large population, sampling without replacement is approximately the same as sampling with replacement, since the odds of choosing the same individual twice is low.

An unbiased random selection of individuals is important so that if a large number of samples were drawn, the average sample would accurately represent the population. However, this does not guarantee that a particular sample is a perfect representation of the population. Simple random sampling merely allows one to draw externally valid conclusions about the entire population based on the sample.

Conceptually, simple random sampling is the simplest of the probability sampling techniques. It requires a complete sampling frame, which may not be available or feasible to construct for large populations. Even if a complete frame is available, more efficient approaches may be possible if other useful information is available about the units in the population.

Advantages are that it is free of classification error, and it requires minimum advance knowledge of the population other than the frame. Its simplicity also makes it relatively easy to interpret data collected in this manner. For these reasons, simple random sampling best suits situations where not much information is available about the population and data collection can be efficiently conducted on randomly distributed items, or where the cost of sampling is small enough to make efficiency less important than simplicity. If these conditions do not hold, stratified sampling or cluster sampling may be a better choice.

Algorithms

Several efficient algorithms for simple random sampling have been developed. A naive algorithm is the draw-by-draw algorithm where at each step we remove the item at that step from the set with equal probability and put the item in the sample. We continue until we have sample of desired size k. The drawback of this method is that it requires random access in the set.

The selection-rejection algorithm developed by Fan et al. in 1962 requires single pass over data; however, it is a sequential algorithm and requires knowledge of total count of items n, which is not available in streaming scenarios.

A very simple random sort algorithm was proved by Sunter in 1977 which simply assigns a random number drawn from uniform distribution (0, 1) as key to each item, sorts all items using the key and selects the smallest k items.

J. Vitter in 1985 proposed reservoir sampling algorithm which is often widely used. This algorithm does not require advance knowledge of n and uses constant space.

Random sampling can also be accelerated by sampling from the distribution of gaps between samples, and skipping over the gaps.

Distinction Between a Systematic Random Sample and a Simple Random Sample

Consider a school with 1000 students, and suppose that a researcher wants to select 100 of them for further study. All their names might be put in a bucket and then 100 names might be pulled out. Not only does each person have an equal chance of being selected, we can also easily calculate the probability *P* of a given person being chosen, since we know the sample size (*n*) and the population (*N*):

1. In the case that any given person can only be selected once (i.e., after selection a person is removed from the selection pool):

$$P = 1 - \frac{N-1}{N} \cdot \frac{N-2}{N-1} \cdot \cdots \cdot \frac{N-n}{N-(n-1)}$$

$$\text{Canceling:} = 1 - \frac{N-n}{N}$$

$$= \frac{n}{N}$$

$$= \frac{100}{1000}$$

$$= 10\%$$

2. In the case that any selected person is returned to the selection pool (i.e., can be picked more than once):

$$P = 1 - \left(1 - \frac{1}{N}\right)^n = 1 - \left(\frac{999}{1000}\right)^{100} = 0.0952\ldots \approx 9.5\%$$

This means that every student in the school has in any case approximately a 1 in 10 chance of being selected using this method. Further, all combinations of 100 students have the same probability of selection.

If a systematic pattern is introduced into random sampling, it is referred to as "systematic (random) sampling". An example would be if the students in the school had numbers attached to their names ranging from 0001 to 1000, and we chose a random starting point, e.g. 0533, and then picked every 10th name thereafter to give us our sample of 100 (starting over with 0003 after reaching 0993). In this sense, this technique is similar to cluster sampling, since the choice of the first unit will determine the remainder. This is no longer simple random sampling, because some combinations of 100 students have a larger selection probability than others – for instance, {3, 13, 23, ..., 993} has a 1/10 chance of selection, while {1, 2, 3, ..., 100} cannot be selected under this method.

Sampling a Dichotomous Population

If the members of the population come in three kinds, say "blue" "red" and "black", the number of red elements in a sample of given size will vary by sample and hence is a random variable whose distribution can be studied. That distribution depends on the numbers of red and black elements in the full population. For a simple random sample *with* replacement, the distribution is a *binomial distribution*. For a simple random sample *without* replacement, one obtains a *hypergeometric distribution*.

- Multistage sampling
- Nonprobability sampling
- Opinion poll
- Quantitative marketing research

Random Sampling Selection Procedure

Suppose there are N units in the population out of which n units are to be selected.

1. Identify the N units in the population with the numbers 1 to N.
2. Choose any random number arbitrarily from the random numbers table and start reading numbers.

3. Choose the sampling unit whose serial number corresponds to the random number drawn from the table of random numbers.

4. In case of SRSWR, all the random numbers are accepted even if repeated more than once.

5. In case of SRSWOR, if any random number is repeated, then it is ignored and more numbers are drawn.

Such process can be implemented through programming and using the discrete uniform distribution. Any number between 1 and N can be generated from this distribution and corresponding unit can be seleced into the sample by associating an index with each sampling unit. Many statistical softwares like R, SAS, etc. have inbuilt functions for drawing a sample using SRSWOR or SRSWR.

Notations

The following notations will be used:

N : Number of sampling units in the population (Population size).

n : Number of sampling units in the sample (Sample size)

Y : The characteristic under consideration

Y_i : Value of the characteristic for the ith unit of the population

$$\bar{Y} = \frac{1}{N}\sum_{i=1}^{N} Y_i \quad : \text{ population mean}$$

$$\bar{y} = \frac{1}{n}\sum_{i=1}^{n} y_i \quad : \text{ sample mean}$$

$$S^2 = \frac{1}{N-1}\sum_{i=1}^{N}(Y_i - \bar{Y})^2 = \frac{1}{N-1}\left(\sum_{i=1}^{N} Y_i^2 - N\bar{Y}^2\right)$$

$$\sigma^2 = \frac{1}{N}\sum_{i=1}^{N}(Y_i - \bar{Y})^2 = \frac{1}{N}\left(\sum_{i=1}^{N} Y_i^2 - N\bar{Y}^2\right)$$

$$s^2 = \frac{1}{n-1}\sum_{i=1}^{n}(y_i - \bar{y})^2 = \frac{1}{n-1}\left(\sum_{i=1}^{n} y_i^2 - n\bar{y}^2\right)$$

Probability of Drawing a Sample

SRSWOR

If n units are selected by SRSWOR, the total number of possible samples are : $\binom{N}{n}$

So the probability of selecting any one of these samples is: $\frac{1}{\binom{N}{n}}$

Alternatively $u_1, u_2, ..., u_n$ are the units selected in the sample, then

$$P(u_1, u_2, ..., u_n) = P(u_1), P(u_2), ..., P(u_n)$$

To compute this expression, consider the probability that a specified unit, say i[th] unit u_i is included in the sample.

The ith unit can be selected either at first draw, second draw, ..., or n[th] draw. Thus the required probability is

$$P_1(i) + P_2(i) + ... + P_n(i)$$

$$= \frac{1}{N} + \frac{1}{N} + ... + \frac{1}{N} (\text{n times})$$

—

Where $P_j(i)$ denotes the probability of selection of u_i at j^{th} draw $j = 1, 2, ..., n$

$$\text{If } P(u_1) = \frac{n}{N}, \text{ then}$$

$$P(u_2) = \frac{n-1}{N-1},$$

.

.

.,

$$P(u_n) = \frac{1}{N-n+1},$$

Thus

$$P(u_1, u_2, ..., u_n) = \frac{n}{N} \cdot \frac{n-1}{N-1} \cdot \frac{n-2}{N-2} \cdots \frac{1}{N-n+1}$$

$$= \frac{1}{\binom{N}{n}}$$

SRSWR

When n units are selected with SRSWR, the total number of possible samples are N^n. The probability of drawing a sample is $\frac{1}{N^n}$.

Alternatively,

$$P(u_1, u_2, ..., u_n) = P(u_1).P(u_2)...., P(u_n)$$

$$= \frac{1}{N} \cdot \frac{1}{N} \cdots \frac{1}{N}$$

$$= \frac{1}{N^n}$$

Probability of Drawing a Unit

SRSWOR

Let A_l denotes an event that a particular unit u_j is not selected at the lth draw. The probability of selecting, say, j^{th} unit at k^{th} draw is

$$
\begin{aligned}
P(\text{selection of } u_j \text{ at } k^{th} \text{ draw}) &= P\left(A_1 \cap A_2 \cap \ldots \cap A_{k-1} \cap \bar{A}_k\right) \\
&= P(A_1)P(A_2/A_1)P(A_3/A_1A_2)\ldots P(A_{k-1}/A_1,A_2\ldots A_{k-2})P\left(\bar{A}_k/A_1,A_2\ldots A_{k-1}\right) \\
&= \left(1-\frac{1}{N}\right)\left(1-\frac{1}{N-1}\right)\left(1-\frac{1}{N-2}\right)\ldots\left(1-\frac{1}{N-k+2}\right)\frac{1}{N-k+1} \\
&= \frac{N-1}{N}.\frac{N-2}{N-1}\ldots\frac{N-k+1}{N-k+2}.\frac{1}{N-k+1} \\
&= \frac{1}{N}
\end{aligned}
$$

SRSWR

$$P(\text{selection of } u_j \text{ at } k^{th} \text{ draw}) = \frac{1}{N}$$

Estimation of Population Mean and Population Variance

One of the main objectives after the selection of a sample is to know about the tendency of the data to cluster around the central value and the scatterdness of the data around the central value.

Among various indicators of central tendency and dispersion, the popular choices are arithmetic mean and variance. So the population mean and population variability are generally measured by arithmetic mean (or weighted arithmetic mean) and variance.

There are various popular estimators for estimating the population mean and population variance. Among them, sample arithmetic mean and sample variance are more popular than other estimators.

One of the reason to use these estimators is that they possess nice statistical properties. Moreover, they are also obtained through well established statistical estimation procedures like maximum likelihood estimation, least squares estimation, method of moments etc. under several standard statistical distributions.

One may also consider other indicators like median, mode, geometric mean, harmonic mean for measuring the central tendency and mean deviation, absolute deviation, Pitman nearness etc. for measuring the dispersion. The properties of such estimators can be studied by numerical procedures like bootstraping.

Estimation of Population Mean

Let us consider the sample arithmetic mean $\bar{y}=\frac{1}{n}\sum_{i=1}^{n}y_i$ as an estimator of population mean $\bar{Y}=\frac{1}{N}\sum_{i=1}^{N}Y_i$ and verify if $\bar{y}$ is an unbiased estimator of $\bar{Y}$ under the two cases.

SRSWOR

$$E(\bar{y})=\frac{1}{n}E\left(\sum_{i=1}^{n}y_i\right)=\frac{1}{n}E(t_i)=\frac{1}{n}\left(\frac{1}{\binom{N}{n}}\sum_{i=1}^{\binom{N}{n}}t_i\right)\left(\text{where } t_i=\sum_{i=1}^{n}y_i\right)$$

$$=\frac{1}{n}\cdot\frac{1}{\binom{N}{n}}\sum_{i=1}^{\binom{N}{n}}\left(\sum_{i=1}^{n}y_i\right).$$

When n units are sampled from N units by without replacement, then each unit of the population can occur with $(n-1)$ other units selected out of the remaining $(N-1)$ units in the population and each unit occurs in $\binom{N-1}{n-1}$ of the $\binom{N}{n}$ possible samples. So

so $\sum_{i=1}^{\binom{N}{n}}\left(\sum_{i=1}^{n}y_i\right)=\binom{N-1}{n-1}\sum_{i=1}^{N}y_i$

Now $E(\bar{y})=\frac{(N-1)!}{(n-1)!(N-n)!}\cdot\frac{n!(N-n)!}{nN!}\sum_{i=1}^{N}y_i=\frac{1}{N}\sum_{i=1}^{N}y_i=\bar{Y}.$

Thus $\bar{y}$ is an unbiased estimator of $\bar{Y}$.

Alternatively, the following approach can also be adopted to show that the sample mean is an unbiased estimator of population mean

$$E(\bar{y})=\frac{1}{n}\sum_{j=1}^{n}E(y_j)$$

$$=\frac{1}{n}\sum_{j=1}^{n}\left[\sum_{i=1}^{N}Y_iP_j(i)\right]$$

$$=\frac{1}{n}\sum_{j=1}^{n}\left[\sum_{i=1}^{N}Y_i.\frac{1}{N}\right]$$

$$=\frac{1}{n}\sum_{j=1}^{n}\bar{Y}=\bar{Y}$$

Where $P_j(i)$ denotes the probability of selection of ith unit at j^{th} stage.

SRSWR

$$E(\bar{y})=\frac{1}{n}E\left(\sum_{i=1}^{n}y_i\right)$$

$$=\frac{1}{n}\sum_{i=1}^{n}E(y_i)=\frac{1}{n}\sum_{i=1}^{n}(Y_1P_1+\ldots+Y_NP_N)=\frac{1}{n}\sum_{i=1}^{n}\bar{Y}$$

$$=\bar{Y}$$

Where $P_i = \frac{1}{N}$ for all $i = 1,2,.....,N$ is the probability of selection of a unit. Thus $\bar{y}$ is an unbiased estimator of population mean under SRSWR also.

Variance of the Estimate

Assume that each observation has same variance σ^2. Then

$$\begin{aligned}
V(\bar{y}) &= E(\bar{y} - \bar{Y})^2 \\
&= E\left[\frac{1}{n}\sum_{i=1}^{n}(y_i - \bar{Y})\right]^2 \\
&= E\left[\frac{1}{n^2}\sum_{i=1}^{n}(y_i - \bar{Y})^2 + \frac{1}{n^2}\sum_{i}^{n}\sum_{\neq j}^{n}(y_i - \bar{Y})(y_j - \bar{Y})\right] \\
&= \frac{1}{n^2}\sum_{i=1}^{n}E(y_i - \bar{Y})^2 + \frac{1}{n^2}\sum_{i}^{n}\sum_{\neq j}^{n}E(y_i - \bar{Y})(y_j - \bar{Y}) \\
&= \frac{1}{n^2}\sum_{i=1}^{n}\sigma^2 + \frac{K}{n^2} \\
&= \frac{N-1}{Nn}S^2 + \frac{K}{n^2}
\end{aligned}$$

Where

$$K = \sum_{i}^{n}\sum_{\neq j}^{n}E(y_i - \bar{Y})(y_j - \bar{Y}).$$

Now we find the value of *K* under the setups of SRSWR and SRSWOR.

SRSWOR

$$K = \sum_{i}^{n}\sum_{\neq j}^{n}E(y_i - \bar{Y})(y_j - \bar{Y}).$$

Consider

$$E(y_i - \bar{Y})(y_j - \bar{Y}) = \frac{1}{N(N-1)}\sum_{k}^{N}\sum_{\neq l}^{N}(y_k - \bar{Y})(y_l - \bar{Y}).$$

Since

$$\left[\sum_{k=1}^{N}(y_k - \bar{Y})\right]^2 = \sum_{k=1}^{N}(y_k - \bar{Y})^2 + \sum_{k}^{n}\sum_{\neq l}^{n}(y_k - \bar{Y})(y_l - \bar{Y})$$

$$0 = (N-1)S^2 + \sum_{k}^{N}\sum_{\neq l}^{N}(y_k - \bar{Y})(y_l - \bar{Y})$$

$$\begin{aligned}
\sum_{k}^{N}\sum_{\neq l}^{N}(y_k - \bar{Y})(y_l - \bar{Y}) &= \frac{1}{N(N-1)}\left[-(N-1)S^2\right] \\
&= -\frac{S^2}{N}
\end{aligned}$$

Thus

$$K = -n(n-1)\frac{S^2}{N}$$

and substituting the value of K, the variance of $\bar{y}$ under SRSWOR is

$$V(\bar{y}_{WOR}) = \frac{N\quad 1}{Nn}S \ -\frac{1}{n}n(n-1)\frac{S}{N}$$
$$\frac{N\quad n}{Nn}S\,.$$

SRSWR

Now we obtain the value of *K* under SRSWR

$$\begin{aligned} K &= \sum_{i}^{n}\sum_{\neq j}^{n} E(y_i - \bar{Y})(y_j - \bar{Y}) \\ &= \sum_{i}^{n}\sum_{\neq j}^{n} E(y_i - \bar{Y})E(y_j - \bar{Y}) \\ &= 0 \end{aligned}$$

because i^{th} and j^{th} draws $(i \neq j)$ are independent. Thus the variance of $\bar{y}$ under SRSWR is

$$V(\bar{y}_{WR}) = \frac{N-1}{Nn}S^2$$

It is to be noted that if N is infinite (large enough), then

$$V(\bar{y}) = \frac{S^2}{n}$$

in both the cases of SRSWOR and SRSWR. So the factor $\frac{N-n}{N}$ is responsible for changing the variance of $\bar{y}$ when the sample is drawn from a finite population in comparison to an infinite population. This is why $\frac{N-n}{N}$ is called as finite population correction (fpc).

It may be noted that $\frac{N-n}{N} = 1 - \frac{n}{N}$ so $\frac{N-n}{N}$ is close to 1 if the ratio of sample size to population size $\left(\frac{n}{N}\right)$ is very small or negligible. In such a case, the size of the population has no direct effect on the variance of $\bar{y}$.

The term $\frac{n}{N}$ is called as sampling fraction.

In practice, fpc can be ignored whenever $\frac{n}{N} < 5\%$ and for many purposes even if it is as high as 10%.

Ignoring fpc will result in the over estimation of variance of $\bar{y}$.

Efficiency of $\bar{y}$ Under SRSWOR over SRSWR:

Now we compare the variances of sample means under SRSWOR and SRSWR.

$$V(\bar{y})_{WOR} = \frac{N-n}{Nn} S^2$$

$$\begin{aligned} V(\bar{y})_{WR} &= \frac{N-1}{Nn} S^2 \\ &= \frac{N-n}{Nn} S^2 + \frac{n-1}{Nn} S^2 \\ &= V(\bar{y})_{WOR} + \text{a positive quantity} \end{aligned}$$

$$V(\bar{y})_{WR} > V(\bar{y})_{WOR}$$

and so, SRSWOR is more efficient than SRSWR.

Estimation of variance from a sample

Since the expressions of variances of sample mean involve S^2 which is based on population values, so these expressions can not be used in real life applications. In order to estimate the variance of $\bar{y}$ on the basis of a sample, an estimator of S^2 (or equivalently σ^2) is needed. Consider s^2 as an estimator of S^2 (or σ^2) and we investigate its biasedness for S^2 in the cases of SRSWOR and SRSWR

Consider

$$\begin{aligned} s^2 &= \frac{1}{n-1}\sum_{i=1}^{n}(y_i - \bar{y})^2 \\ &= \frac{1}{n-1}\sum_{i=1}^{n}\left[(y_i - \bar{Y}) - (\bar{y} - \bar{Y})\right]^2 \\ &= \frac{1}{n-1}\left[\sum_{i=1}^{n}(y_i - \bar{Y})^2 - n(\bar{y} - \bar{Y})^2\right] \end{aligned}$$

$$\begin{aligned} E(s^2) &= \frac{1}{n-1}\left[\sum_{i=1}^{n}E(y_i - \bar{Y})^2 - nE(\bar{y} - \bar{Y})^2\right] \\ &= \frac{1}{n-1}\left[\sum_{i=1}^{n}Var(y_i) - n\,Var(\bar{y})\right] \\ &= \frac{1}{n-1}\left[n\sigma^2 - n\,Var(\bar{y})\right] \end{aligned}$$

In case of SRSWOR

$$Var(\bar{y})_{WOR} = \frac{N-n}{Nn} S^2$$

$$\begin{aligned} \text{and so } E(s^2) &= \frac{n}{n-1}\left[\sigma^2 - \frac{N-n}{Nn} S^2\right] \\ &= \frac{n}{n-1}\left[\frac{N-1}{N} S^2 - \frac{N-n}{Nn} S^2\right] \\ &= S^2 \end{aligned}$$

In case of SRSWR

$$Var(\bar{y})_{WR} = \frac{N-1}{Nn}S^2$$

$$\text{and so} \quad E(s^2) = \frac{n}{n-1}\left[\sigma^2 - \frac{N-1}{Nn}S^2\right]$$
$$= \frac{n}{n-1}\left[\frac{N-1}{N}S^2 - \frac{N-1}{Nn}S^2\right]$$
$$= \frac{N-1}{N}S^2$$
$$= \sigma^2$$

$$\text{Hence } E(s^2) = \begin{cases} S^2 & \text{in SRSWOR} \\ \sigma^2 & \text{in SRSWR} \end{cases}$$

An unbiased estimate of $Var(\bar{y})$ is

$\widehat{Var}(\bar{y})_{WOR} = \frac{N-n}{Nn}s^2$ in case of SRSWOR.

$$\widehat{Var}(\bar{y})_{WR} = \frac{N-1}{Nn}.\frac{N}{N-1}s^2 = \frac{s^2}{n}$$ in case of SRSWR.

Standard Errors

The standard error of $\bar{y}$ is defined as $\sqrt{Var(\bar{y})}$

In order to estimate the standard error, one simple option is to consider the square root of estimate of variance of sample mean.

- under SRSWOR, a possible estimator is

$$\hat{\sigma}(\bar{y}) = \sqrt{\frac{N-n}{Nn}}s$$

- under SRSWR, a possible estimator is

$$\hat{\sigma}(\bar{y}) = \sqrt{\frac{N-1}{Nn}}s$$

It is to be noted that this estimator does not possess the same properties as of $\widehat{Var}(\bar{y})$

Reason being if $\hat{\theta}$ is an estimator of θ, then $\widehat{\sqrt{\theta}}$ is not necessarily an estimator of

In fact, the $\hat{\sigma}(\bar{y})$ is a negatively biased estimator under SRSWOR.

Consider s as an estimator of S.

Let

$$s^2 = S^2 + \varepsilon \text{ with } E(\varepsilon^2) = S^2$$

Write

$$\begin{aligned} s &= (S^2 + \varepsilon)^{1/2} \\ &= S\left(1 + \frac{\varepsilon}{S^2}\right)^{1/2} \\ &= S\left(1 + \frac{\varepsilon}{2S^2} - \frac{\varepsilon^2}{8S^4} + \ldots\right) \end{aligned}$$

Assuming ε will be small as compared to S^2 and as n becomes large, the probability of such an event approaches one.

Neglecting the powers of ε higher than two and taking expectation, we have

$$E(s) \simeq \left[1 - \frac{Var(s^2)}{8S^4}\right] S$$

Where

$$Var(s^2) = \frac{2S^4}{(n-1)}\left[1 + \left(\frac{n-1}{2n}\right)(\beta_2 - 3)\right] \text{for large N.}$$

$$\mu_j = \frac{1}{N}\sum_{i=1}^{N}(Y_i - \bar{Y})^j$$

$$\beta_2 = \frac{\mu_4}{S^4} : \text{coefficient of kurtosis}$$

Thus

$$\begin{aligned} E(s) &\simeq S\left[1 - \frac{1}{4(n-1)} - \frac{\beta_2 - 3}{8n}\right] \\ Var(s) &\simeq S^2 - S^2\left[1 - \frac{1}{8}\frac{Var(s^2)}{S^4}\right]^2 \\ &= \frac{Var(s^2)}{4S^2} \\ &= \frac{S^2}{2(n-1)}\left[1 + \left(\frac{n-1}{2n}\right)(\beta_2 - 3)\right]. \end{aligned}$$

Note that for a normal distribution, $\beta_2 = 3$ and we obtain

$$Var(s) = \frac{S^2}{2(n-1)}$$

Both $Var(s)$ and $Var(s^2)$ are inflated due to nonnormality to the same extent, by the inflation factor

$$\left[1+\left(\frac{n-1}{2n}\right)(\beta_2-3)\right]$$

and this does not depends on coefficient of skewness.

This is an important result to be kept in mind while determining the sample size in which it is assumed that S^2 is known. If inflation factor is ignored and population is non-normal, then the reliability on s^2 may be misleading.

Alternative Approach

The results for the unbiasedness property and variance of sample mean can also be proved in an alternative way as follows:

(i) SRSWOR

With the ith unit of the population, we associate a random variable a_i defined as follows:

$$a_i = \begin{cases} 1 & \text{if the } i^{th} \text{ unit occurs in the sample} \\ 0 & \text{if the } i^{th} \text{ unit does not occur in the sample } (i=1,2,....N). \end{cases}$$

Then,

$$E(a_i) = 1 \times \text{Probability that the } i^{th} \text{ unit is included in the sample}$$

$$= \frac{n}{N}, i = 1,2,....,N.$$

$$E(a_i^2) = 1 \times \text{Probability that the } i^{th} \text{ unit is included in the sample}$$

$$= \frac{n}{N}, i = 1,2,....,N.$$

$$E(a_i a_j) = 1 \times \text{Probability that the } i^{th} \text{and } j^{th} \text{unit is included in the sample}$$

$$= \frac{n(n-1)}{N(N-1)}, i \neq j = 1,2,....,N.$$

From these results, we can obtain

$$Var(a_i) = E(a_i^2) - (E(a_i))^2 = \frac{n(N-n)}{N^2}, i = 1,2,.....,N$$

$$Cov(a_i, a_j) = E(a_i a_j) - E(a_i)E(a_j) = \frac{n(N-n)}{N^2(N-1)}, i \neq j = 1,2,.....,N.$$

We can rewrite the sample mean as

$$\bar{y} = \frac{1}{n}\sum_{i=1}^{N} a_i y_i$$

Then

$$E(\bar{y}) = \frac{1}{n}\sum_{i=1}^{N} E(a_i) y_i = \bar{Y}$$

and

$$Var(\bar{y}) = \frac{1}{n^2} Var\left(\sum_{i=1}^{N} a_i y_i\right) = \frac{1}{n^2}\left[\sum_{i=1}^{N} Var(a_i) y_i^2 + \sum_{i \neq j}^{N} Cov(a_i, a_j) y_i y_j\right].$$

Substituting the values of $Var(a_i)$ and $Cov(a_i, a_j)$ in the expression of $Var(\bar{y})$ and simplifying, we get

$$Var(\bar{y}) = \frac{N-n}{Nn} S^2.$$

To show that $E(s^2) = S^2$, consider

$$s^2 = \frac{1}{(n-1)}\left[\sum_{i=1}^{n} y_i^2 - n\bar{y}^2\right] = \frac{1}{(n-1)}\left[\sum_{i=1}^{N} a_i y_i^2 - n\bar{y}^2\right]$$

Hence, taking, expectation, we get

$$E(s^2) = \frac{1}{(n-1)}\left[\sum_{i=1}^{N} E(a_i) y_i^2 - n\left\{Var(\bar{y}) + \bar{Y}^2\right\}\right]$$

Substituting the values of $E(a_i)$ and $Var(\bar{y})$ in this expression and simplifying we get $E(s^2) = S^2$.

(ii) SRSWR

Let a random variable a_i associated with the ith unit of the population denotes the number of times the ith unit occurs in the sample $i = 1, 2, \ldots, N$. So a_i assumes values 0, 1, 2,...,n. The joint distribution of $a_1, a_2, \ldots, a_N$ is the multinomial distribution given by

$$P(a_1, a_2, \ldots, a_N) = \frac{n!}{\prod_{i=1}^{N} a_i!} \cdot \frac{1}{N^n}$$

Where $\sum_{i=1}^{N} a_i = n$. For this multinomial distribution, we have

$$E(a_i) = \frac{n}{N}$$

$$Var(a_i) = \frac{n(N-1)}{N^2}, i = 1, 2, ..., N.$$

$$Cov(a_i, a_j) = -\frac{n}{N^2}, \ i \neq j = 1, 2, ..., N.$$

We rewrite the sample mean as

$$\bar{y} = \frac{1}{n}\sum_{i=1}^{N} a_i y_i.$$

Hence, taking expectation of $\bar{y}$ and substituting the value of $E(a_i) = n/N$, we obtain that $E(\bar{y}) = \bar{Y}$.

Further

$$Var(\bar{y}) = \frac{1}{n^2}\left[\sum_{i=1}^{N} Var(a_i) y_i^2 + \sum_{i=1}^{N} Cov(a_i, a_j) y_i y_j\right].$$

Substituting, the values of $Var(a_i) = n(N-1)/N^2$ and $Cov(a_i, a_j) = -n/N^2$ and simplifying, we get

$$Var(\bar{y}) = \frac{N-1}{Nn} S^2.$$

To prove that $E(s^2) = \frac{N-1}{N} S^2 = \sigma^2$ in SRSWR, consider

$$(n-1)s^2 = \sum_{i=1}^{n} y_i^2 - n\bar{y}^2 = \sum_{i=1}^{N} a_i y_i^2 - n\bar{y}^2,$$

$$(n-1)E(s^2) = \sum_{i=1}^{N} E(a_i) y_i^2 - n\left\{Var(\bar{y}) + \bar{Y}^2\right\}$$

$$= \frac{n}{N}\sum_{i=1}^{N} y_i^2 - n.\frac{(N-1)}{nN} S^2 - n\bar{Y}^2$$

$$= \frac{(n-1)(N-1)}{N} S^2$$

$$E(s^2) = \frac{N-1}{N} S^2 = \sigma^2$$

Estimator of Population Total

Sometimes, it is also of interest to estimate the population total, e.g. total household income, total expenditures etc. Let Y_T denotes the population total

$$Y_T = \sum_{i=1}^{N} Y_i = N\bar{Y}$$

which can be estimated by

$$\hat{Y}_T = N\hat{\bar{Y}} = N\bar{y}$$

Obviously

$$E(\hat{Y}_T) = NE(\bar{y}) = N\bar{Y} = Y_T$$

$$Var(\hat{Y}_T) = N^2 Var(\bar{y}) = \begin{cases} N^2\left(\frac{N-n}{Nn}\right)S^2 = \frac{N(N-n)}{n}S^2 & \text{for SRSWOR} \\ N^2\left(\frac{N-1}{Nn}\right)S^2 = \frac{N(N-1)}{n}S^2 & \text{for SRSWR} \end{cases}$$

and the estimates of variance of $\hat{Y}_T$ are

$$\widehat{Var}(\hat{Y}_T) = \begin{cases} \frac{N(N-n)}{n}s^2 & \text{for SRSWOR} \\ \frac{N^2}{n}s^2 & \text{for SRSWR} \end{cases}$$

Confidence Limits for the Population Mean

Now we construct the $100(1-\alpha)\%$ confidence interval for the population mean.

Assume that the population is normally distributed $N(\mu,\sigma^2)$ with mean μ and variance σ^2, then $\frac{\bar{y}-\bar{Y}}{\sqrt{Var(\bar{y})}}$ follows N(0,1). when σ^2 is known.

If σ^2 is unknown and is estimated from the sample then $\frac{\bar{y}-\bar{Y}}{\sqrt{Var(\bar{y})}}$ follows a t-distribution with (n -1) degrees of freedom.

When σ^2 is known, then the $100(1-\alpha)\%$ confidence interval is given by

$$P\left[-Z_{\frac{\alpha}{2}} \le \frac{\bar{y}-\bar{Y}}{\sqrt{Var(\bar{y})}} \le Z_{\frac{\alpha}{2}}\right] = 1-\alpha$$

$$\text{or } P\left[\bar{y} - Z_{\frac{\alpha}{2}}\sqrt{Var(\bar{y})} \le \bar{Y} \le \bar{y} + Z_{\frac{\alpha}{2}}\sqrt{Var(\bar{y})}\right] = 1-\alpha$$

and the confidence limits are

$$\left(\bar{y} - Z_{\frac{\alpha}{2}}\sqrt{Var(\bar{y})}, \bar{y} + Z_{\frac{\alpha}{2}}\sqrt{Var(\bar{y})}\right)$$

where $Z_{\frac{\alpha}{2}}$ denotes the upper $\frac{\alpha}{2}$% points on $N(0,1)$ distribution.

Similarly, when σ^2 is unknown, then the $100(1-\alpha)\%$ confidence interval is

$$P\left[-t_{\frac{\alpha}{2}} \le \frac{\bar{y}-\bar{Y}}{\sqrt{\widehat{Var}(\bar{y})}} \le t_{\frac{\alpha}{2}}\right] = 1-\alpha$$

$$\text{or } P\left[\bar{y}-t_{\frac{\alpha}{2}} \le \sqrt{\widehat{Var}(\bar{y})} \le \bar{Y} \le \bar{y}+t_{\frac{\alpha}{2}}\sqrt{\widehat{Var}(\bar{y})}\right] = 1-\alpha$$

and the confidence limits are

$$\left[\bar{y}-t_{\frac{\alpha}{2}} \le \sqrt{\widehat{Var}(\bar{y})} \le \bar{Y} \le \bar{y}+t_{\frac{\alpha}{2}}\sqrt{\widehat{Var}(\bar{y})}\right]$$

where $t_{\frac{\alpha}{2}}$ denotes the upper $\frac{\alpha}{2}$ % points of t-distribution with (n - 1) degrees of freedom

Determination of Sample Size

The size of the sample is needed before the survey starts and goes into operation. One point to be kept in mind is that when the sample size increases, the variance of estimators decreases but the cost of survey increases and vice versa.

So there has to be a balance between the two aspects. The sample size can be determined on the basis of prescribed values of standard error of sample mean, error of estimation, width of the confidence interval, coefficient of variation of sample mean, relative error of sample mean or total cost among several others.

An important constraint or need to determine the sample size is that the information regarding the population standard derivation S should be known for these criterion. The reason and need for this will be clear when we derive the sample size. A question arises about how to have information about S before hand? The possible solutions to this issue is to conduct a pilot survey and collect a preliminary sample of small size, estimate S and use it as known value of S. Alternatively, such information can also be collected from past data, past experience, long association of experimenter with the experiment, prior information, etc.

Now we find the sample size under different criteria assuming that the samples have been drawn using SRSWOR. The sample sizes under SRSWR can be derived similarly.

Pre-specified Variance

The sample size is to be determined such that the variance of sample mean should not exceed a given value, say V. In this case, find n such that

$$\begin{aligned} & Var(\bar{y}) \leq V \\ \text{or} \quad & \frac{N-n}{Nn} S^2 \leq V \\ \text{or} \quad & \frac{1}{n} - \frac{1}{N} \leq \frac{V}{S^2} \\ \text{or} \quad & \frac{1}{n} - \frac{1}{N} \leq \frac{1}{n_1} \\ \text{or} \quad & n \geq \frac{n_1}{1 + \frac{n_1}{N}} \end{aligned}$$

where

$$n_1 = \frac{S^2}{V}$$

It may be noted here that n_1 can be known only when S^2 is known. This reason compels to assume that S should be known. The same reason will also be seen in other cases.

The smallest sample size needed in this case is

$$n_{smallest} = \frac{n_1}{1 + \frac{n_1}{N}}$$

If *N* is large, then the required *n* is $n \geq n_1$ and $n_{smallest} = n_1$

Pre-specified estimation error

It may be possible to have some prior knowledge of population mean $\bar{Y}$ and it may be required that the sample mean $\bar{y}$ should not differ from it by more than a specified amount of absolute estimation error e, which is a small quantity. Such requirement can be satisfied by associating a probability $(1-\alpha)$ with it and can be expressed as

$$P\left[\left|\bar{y} - \bar{Y}\right| \leq e\right] = (1-\alpha).$$

Since $\bar{y}$ follows $N\left(\bar{Y}, \frac{N-n}{Nn} S^2\right)$ assuming the normal distribution for the population, we can write

$$P\left[\frac{\left|\bar{y} - \bar{Y}\right|}{\sqrt{Var(\bar{y})}} \leq \frac{e}{\sqrt{Var(\bar{y})}}\right] = 1 - \alpha$$

which implies that

$$\frac{e}{\sqrt{Var(\bar{y})}} = Z_{\frac{\alpha}{2}}$$

$$\text{or} \quad Z^2_{\frac{\alpha}{2}} Var(\bar{y}) = e^2$$

$$\text{or} \quad Z^2_{\frac{\alpha}{2}} \frac{N-n}{Nn} S^2 = e^2$$

$$\text{or} \quad n = \left(\frac{\left(\frac{Z_{\frac{\alpha}{2}} S}{e}\right)^2}{1 + \frac{1}{N}\left(\frac{Z_{\frac{\alpha}{2}} S}{e}\right)^2} \right)$$

which is the required sample size.

If N is large then $n = \left(\frac{Z_{\frac{\alpha}{2}} S}{e}\right)^2$.

Pre-specified Width of Confidence Interval

If the requirement is that the width of the confidence interval of $\bar{y}$ with confidence co-efficient $(1-\alpha)$ should not exceed a prespecified amount W, then the sample size n is determined such that

$$2Z_{\frac{\alpha}{2}} \sqrt{Var(\bar{y})} \leq W$$

Assuming σ^2 is known and population is normally distributed. This can be expressed as

$$2Z_{\frac{\alpha}{2}} \sqrt{\frac{N-n}{Nn}} S \leq W$$

$$\text{or} \quad 4Z^2_{\frac{\alpha}{2}} \left(\frac{1}{n} - \frac{1}{N}\right) S^2 \leq W^2$$

$$\text{or} \quad \frac{1}{n} \leq \frac{1}{N} + \frac{W^2}{4Z^2_{\frac{\alpha}{2}} S^2}$$

$$\text{or} \quad n \geq \left(\frac{\left(\frac{4Z^2_{\frac{\alpha}{2}} S^2}{W^2}\right)}{1 + \frac{1}{N}\left(\frac{4Z^2_{\frac{\alpha}{2}} S^2}{W^2}\right)} \right)$$

The minimum sample size required is

$$n_{smallest} = \left(\frac{\left(\frac{4Z^2_{\frac{\alpha}{2}} S^2}{W^2} \right)}{1 + \frac{1}{N}\left(\frac{4Z^2_{\frac{\alpha}{2}} S^2}{W^2} \right)} \right)$$

If N is large then

$$n \geq \frac{4Z^2_{\frac{\alpha}{2}} S^2}{W^2}$$

and minimum sample size needed is

$$n_{smallest} = \frac{4Z^2_{\frac{\alpha}{2}} S^2}{W^2}$$

Pre-specified Coefficient of Variation

The coefficient of variation (CV) is defined as the ratio of standard error (or standard deviation) and mean. The knowledge of coefficient of variation has played an important role in the sampling theory as this information helps in deriving efficient estimators of population mean.

If it is desired that the coefficient of variation of $\bar{y}$ should not exceed a given or prespecified value of coefficient of variation, say C_0, then the required sample size n is to be determined such that

$$CV(\bar{y})$$

$$\text{or} \quad \frac{\sqrt{Var(\bar{y})}}{\bar{Y}} \leq C_0$$

$$\text{or} \quad \frac{\left(\frac{N-n}{Nn} S^2 \right)}{\bar{Y}^2} \leq C_0^2$$

$$\text{or} \quad \frac{1}{n} - \frac{1}{N} \leq \frac{C_0^2}{C^2}$$

$$\text{or} \quad n \geq \frac{\frac{C^2}{C_0^2}}{1 + \frac{C^2}{NC_0^2}} \text{ where } C\frac{S}{\bar{Y}} \text{ is the population of variation}$$

The smallest sample size needed in this case is

$$n_{smallest} = \frac{\frac{C^2}{C_0^2}}{1 + \frac{C^2}{NC_0^2}}$$

If N is large, then $n \geq \frac{C^2}{C_0^2}$ and $n_{smallest} = \frac{C^2}{C_0^2}$

Pre-specified relative error

When $\bar{y}$ is used for estimating the population mean $\bar{Y}$, then the relative estimation error is defined as $\frac{\bar{y}-\bar{Y}}{\bar{Y}}$. If it is required that such relative estimation error should not exceed a prespecified value R with probability $(1-\alpha)$, then such requirement can be satisfied by expressing it like

$$P\left[\frac{|\bar{y}-\bar{Y}|}{\sqrt{Var(\bar{y})}} \leq \frac{R\bar{Y}}{\sqrt{Var(\bar{y})}}\right] = 1-\alpha$$

Assuming the population to be normally distributed, $\bar{y}$ follows $N\left(\bar{Y}, \frac{N-n}{Nn}S^2\right)$.

So it can be written that

$$\frac{R\bar{Y}}{\sqrt{Var(\bar{y})}} = Z_{\frac{\alpha}{2}}$$

$$\text{or} \quad Z_{\frac{\alpha}{2}}\left(\frac{N-n}{Nn}\right)S^2 = R^2\bar{Y}^2$$

$$\text{or} \quad \left(\frac{1}{n}-\frac{1}{N}\right) = \frac{R^2}{C^2 Z_{\frac{\alpha}{2}}^2}$$

$$\text{or} \quad n = \frac{\left(\frac{Z_{\frac{\alpha}{2}}C}{R}\right)^2}{1+\frac{1}{N}\left(\frac{Z_{\frac{\alpha}{2}}C}{R}\right)^2}$$ where $C = \frac{S}{\bar{Y}}$ is the population coefficient of variation and should be known

If N is large, then $n = \left(\frac{Z_{\frac{\alpha}{2}}C}{R}\right)^2$

Pre-specified Cost

Let an amount of money C be designated for a sample survey to collect n observations, Further, let C_0 be the overhead cost and C_1 be the cost of collection of one unit in the sample. Then, the total cost C can be expressed as:

$$C = C_0 + nC_1$$

$$\text{or} \quad n = \frac{C - C_0}{C_1}$$

is the required sample size.

In many situations, the characteristic under study on which the observations are collected are qualitative in nature. For example, the responses of customers in many marketing surveys are based on replies like 'yes' or 'no', 'agree' or 'disagree' etc. Sometimes the respondents are asked to order the several options like first choice, second choice etc. Sometimes the objective of the survey is to estimate the proportion or the percentage of brown eyed persons, unemployed persons, graduate persons or persons favoring a proposal, etc. In such situations, the first question arises how to do the sampling and secondly how to estimate the population parameters like population mean, population variance, etc.

Sampling Procedure

The same sampling procedures that are used for drawing a sample in case of quantitative characteristics can also be used for drawing a sample for qualitative characteristic. So, the sampling procedures remain same irrespective of the nature of characteristic under study - either qualitative or quantitative. For example, the SRSWOR and SRSWR procedures for drawing the samples remain the same for qualitative and quantitative characteristics. Similarly, other sampling schemes like stratified sampling, two stage sampling etc. also remain same.

Estimation of Population Proportion

The population proportion in case of qualitative characteristic can be estimated in a similar way as the estimation of population mean in case of quantitative characteristic.

Consider a qualitative characteristic based on which the population can be divided into two mutually exclusive classes, say C and C*. For example, if C is the part of population of persons saying 'yes' or 'agreeing' with the proposal then C* is the part of population of persons saying 'no' or 'disagreeing' with the proposal. Let A be the number of units in C and (N - A) units be in C* in a population of size N. Then the proportion of units in C is $P = \frac{A}{N}$

and the proportion of units in C* is

$$Q = \frac{N-A}{N} = 1 - P.$$

An indicator variable Y can be associated with the characteristic under study and then for i = 1,2,..,N

$$Y_i = \begin{cases} 1 & i^{th} \text{ unit belongs to C} \\ 0 & i^{th} \text{ unit belongs to } C^*. \end{cases}$$

Now the population total is $Y_{TOTAL} \sum_{i=1}^{N} Y_i = A$

and population mean is $\overline{Y} = \frac{\sum_{i=1}^{N} Y_i}{N} = \frac{A}{N} = P$

Suppose a sample of size n is drawn from a population of size N by simple random sampling. Let a be the number of units in the sample which fall into class C and (n - a) units falls in class C*, then the sample proportion of units in C is

$$P = \frac{a}{n}$$

which can be written as

$$p = \frac{a}{n} = \frac{\sum_{i=1}^{n} Y_i}{n} = \overline{y}.$$

Since $\sum_{i=1}^{N} Y_i^2 = A = NP$, so we can write S^2 and s^2 in terms of P and Q as

$$\begin{aligned} S^2 &= \frac{1}{N-1}\sum_{i=1}^{N}(Y_i - \overline{Y})^2 \\ &= \frac{1}{N-1}\left(\sum_{i=1}^{N} Y_i^2 - N\overline{Y}^2\right) \\ &= \frac{1}{N-1}(NP - NP^2) \\ &= \frac{N}{N-1}PQ. \end{aligned}$$

Similarly,

$$\sum_{i=1}^{n} y_i^2 = a = np$$

and

$$\begin{aligned} s^2 &= \frac{1}{n-1}\sum_{i=1}^{n}(y_i - \overline{y})^2 \\ &= \frac{1}{n-1}\left(\sum_{i=1}^{n} y_i^2 - n\overline{y}^2\right) \\ &= \frac{1}{n-1}(np - np^2) \\ &= \frac{n}{n-1}pq. \end{aligned}$$

Note that the quantities $\overline{y}, \overline{Y}, s^2$ and S^2 have been expressed as functions of sample and population proportions. Since the sample has been drawn by simple random sampling and sample proportion is same as the sample mean, so the properties of sample proportion in SRSWOR and SRSWR can be derived using the properties of sample mean directly

SRSWOR

Since sample mean $\bar{y}$ an unbiased estimator of population mean $\bar{Y}$, i.e. $E(\bar{y}) = \bar{Y}$ in case of SRSWOR, so

$$E(p) = E(\bar{y}) = \bar{Y} = P$$

and p is an unbiased estimator of P.

Using the expression of $Var(\bar{y})$, the variance of *p* can be derived as

$$\begin{aligned}Var(p) &= Var(\bar{y}) \\ &= \frac{N-n}{Nn}S^2 \\ &= \frac{N-n}{Nn}.\frac{N}{N-1}PQ \\ &= \frac{N-n}{N-1}.\frac{PQ}{n}.\end{aligned}$$

Similarly, using the estimate of $Var(\bar{y})$, the estimate of variance can be derived as

$$\begin{aligned}\widehat{Var}(p) &= \widehat{Var}(\bar{y}) \\ &= \frac{N-n}{Nn}s^2 \\ &= \frac{N-n}{Nn}\frac{n}{n-1}pq \\ &= \frac{N-n}{N(n-1)}pq.\end{aligned}$$

SRSWR

Since sample mean $\bar{y}$ is an unbiased estimator of population mean $\bar{Y}$ in case of SRS-WR so the sample proportion,

$$E(p) = E(\bar{y}) = \bar{Y} = P$$

i.e. *p* is an unbiased estimator of *P*.

Using the expression of $Var(\bar{y})$ and its estimate in case of SRSWR, the variance of *p* and its estimate can be derived as follows:

$$\begin{aligned}Var(p) = Var(\bar{y}) &= \frac{N-1}{Nn}S^2 \\ &= \frac{N-1}{Nn}\frac{N}{N-1}PQ \\ &= \frac{PQ}{n}\end{aligned}$$

$$\widehat{Var}(p) = \frac{n}{n-1}.\frac{pq}{n}$$
$$= \frac{pq}{n-1}.$$

Estimation of Population Total or Total Number of Count

It is easy to see that an estimate of population total A (or total number of count) is $\hat{A} = Np = \frac{Na}{n}$, its variance is $Var(\hat{A}) = N^2 Var(p)$ and estimate of variance is $\widehat{Var}(\hat{A}) = N^2 \widehat{Var}(p)$.

Confidence Interval Estimation of P

If N and n are large then $\frac{p-P}{\sqrt{Var(p)}}$ approximately follows N(0,1). With this approximation, we can write

$$P\left[-Z_{\frac{\alpha}{2}} \le \frac{p-P}{\sqrt{Var(p)}} \le Z_{\frac{\alpha}{2}}\right] = 1-\alpha$$

and the 100(1–α)% confidence interval of P is

$$\left(p - Z_{\frac{\alpha}{2}}\sqrt{Var(p)}, p + Z_{\frac{\alpha}{2}}\sqrt{Var(p)}\right).$$

It may be noted that in this case, a discrete random variable is being approximated by a continuous random variable, so a continuity correction n/2 can be introduced in the confidence limits and the limits become

$$\left(p - Z_{\frac{\alpha}{2}}\sqrt{Var(p)} + \frac{n}{2}, p + Z_{\frac{\alpha}{2}}\sqrt{Var(p)} - \frac{n}{2}\right).$$

Use of Hypergeometric Distribution

When SRS is applied for the sampling of a qualitative characteristic, the methodology is to draw the units one- by-one and so the probability of selection of every unit remains same at every step. If n sampling units are selected together from N units, then the probability of selection of units does not remain same as in the case of SRS.

Consider a situation in which the sampling units in a population are divided into two mutually exclusive classes. Let P and Q be the proportions of sampling units in the population belonging to classes '1' and '2' respectively. Then NP and NQ are the total number of sampling units in the population belonging to class '1' and '2' respectively and so NP + NQ = N. The probability that in a sample of n selected units out of N units by SRS such that n_1 selected units belongs to class '1' and n_2 selected units belongs to class '2' is governed by the hypergeometric distribution and

$$P(n_1) = \frac{\binom{NP}{n_1}\binom{NQ}{n_2}}{\binom{N}{n}}$$

As N grows large, the hypergeometric distribution tends to Binomial distribution and $P(n_1)$ is approximated by

$$P(n_1) = \binom{n}{n_1} p^{n_1} (1-p)^{n_2}$$

Inverse Transform Sampling

Inverse transform sampling (also known as inversion sampling, the inverse probability integral transform, the inverse transformation method, Smirnov transform, golden rule) is a basic method for pseudo-random number sampling, i.e. for generating sample numbers at random from any probability distribution given its cumulative distribution function.

Inverse transformation sampling takes uniform samples of a number u between 0 and 1, interpreted as a probability, and then returns the largest number x from the domain of the distribution $P(X)$ such that $P(-\infty < X < x) \leq u$. For example, imagine that $P(X)$ is the standard normal distribution with mean zero and standard deviation one. The table below shows samples taken from the uniform distribution and their representation on the standard normal distribution.

Transformation from uniform sample to normal	
u	$F^{-1}(u)$
.5	0
.975	1.95996
0.995	2.5758
.999999	4.75342
1-2^{-52}	8.12589

We are randomly choosing a proportion of the area under the curve and returning the number in the domain such that exactly this proportion of the area occurs to the left of that number. Intuitively, we are unlikely to choose a number in the far end of tails because there is very little area in them which would require choosing a number very close to zero or one.

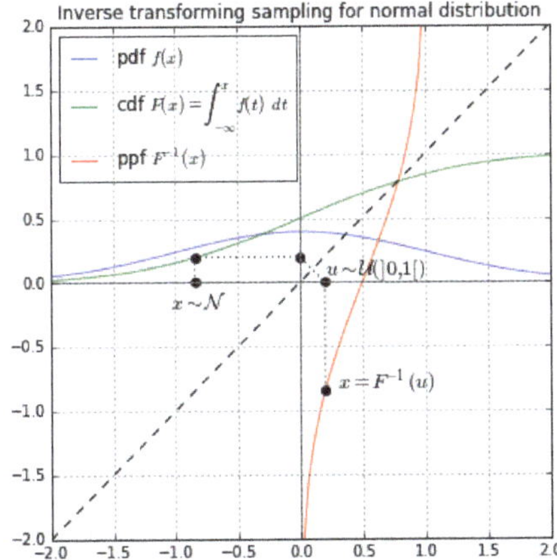

Inverse transform sampling for normal distribution

Computationally, this method involves computing the quantile function of the distribution — in other words, computing the cumulative distribution function (CDF) of the distribution (which maps a number in the domain to a probability between 0 and 1) and then inverting that function. This is the source of the term "inverse" or "inversion" in most of the names for this method. Note that for a discrete distribution, computing the CDF is not in general too difficult: we simply add up the individual probabilities for the various points of the distribution. For a continuous distribution, however, we need to integrate the probability density function (PDF) of the distribution, which is impossible to do analytically for most distributions (including the normal distribution). As a result, this method may be computationally inefficient for many distributions and other methods are preferred; however, it is a useful method for building more generally applicable samplers such as those based on rejection sampling.

For the normal distribution, the lack of an analytical expression for the corresponding quantile function means that other methods (e.g. the Box–Muller transform) may be preferred computationally. It is often the case that, even for simple distributions, the inverse transform sampling method can be improved on: for example, the ziggurat algorithm and rejection sampling. On the other hand, it is possible to approximate the quantile function of the normal distribution extremely accurately using moderate-degree polynomials, and in fact the method of doing this is fast enough that inversion sampling is now the default method for sampling from a normal distribution in the statistical package R.

Definition

The probability integral transform states that if X is a continuous random variable with cumulative distribution function F_X, then the random variable $Y = F_X(X)$ has a uniform distribution on [0, 1]. The inverse probability integral transform is just the inverse of this: specifically, if Y has a uniform distribution on [0, 1] and if X has a cumulative distribution F_X, then the random variable $F_X^{-1}(Y)$ has the same distribution as X.

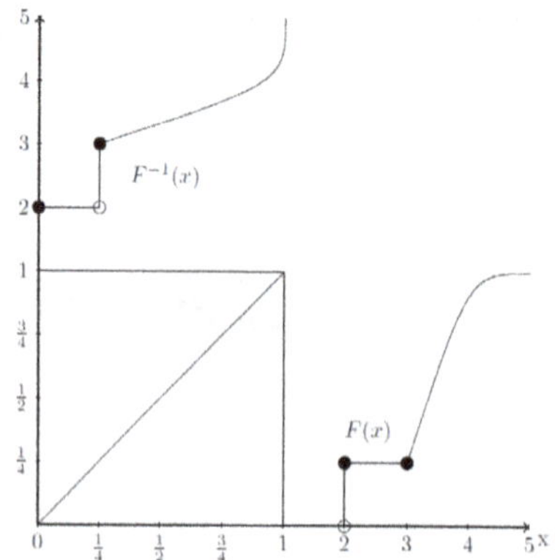

Graph of the inversion technique from x to $F(x)$. On the bottom right we see the regular function and in the top left its inversion.

The Method

The problem that the inverse transform sampling method solves is as follows:

- Let X be a random variable whose distribution can be described by the cumulative distribution function F.
- We want to generate values of X which are distributed according to this distribution.

The inverse transform sampling method works as follows:

1. Generate a random number u from the standard uniform distribution in the interval [0,1].
2. Compute the value x such that $F(x) = u$.
3. Take x to be the random number drawn from the distribution described by F.

Expressed differently, given a continuous uniform variable U in [0, 1] and an invertible cumulative distribution function F, the random variable $X = F^{-1}(U)$ has distribution F (or, X is distributed F).

A treatment of such inverse functions as objects satisfying differential equations can be given. Some such differential equations admit explicit power series solutions, despite their non-linearity.

Examples

- As an example, suppose we have a random variable $U \sim \mathrm{Unif}(0,1)$ and a cumulative distribution function

$$F(x) = 1 - \exp(-\sqrt{x})$$

In order to perform an inversion we want to solve for $F(F^{-1}(u)) = u$

$$\begin{aligned} F(F^{-1}(u)) &= u \\ 1-\exp\left(-\sqrt{F^{-1}(u)}\right) &= u \\ F^{-1}(u) &= (-\log(1-u))^2 \\ &= (\log(1-u))^2 \end{aligned}$$

From here we would perform steps one, two and three.

- As another example, we use the exponential distribution with $F(x) = 1 - e^{-\lambda x}$ for $x \geq 0$ (and 0 otherwise). By solving y=F(x) we obtain the inverse function

$$x = F^{-1}(y) = -\frac{1}{\lambda}\ln(1-y).$$

The idea is illustrated in the following graph:

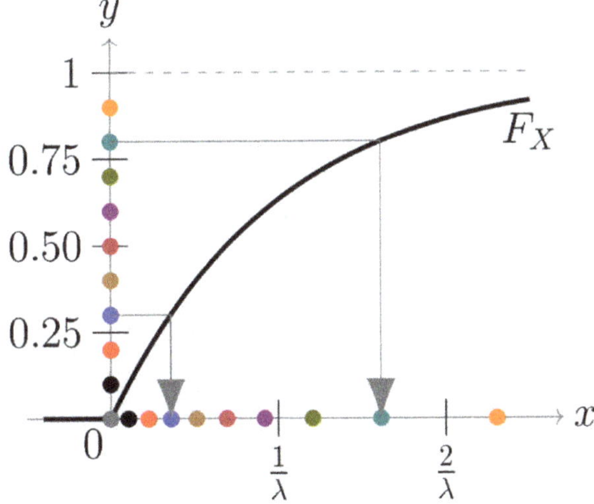

Random numbers y_i are generated from a uniform distribution between 0 and 1, i.e. Y ~ U(0, 1). They are sketched as colored points on the y-axis. Each of the points is mapped according to $x=F^{-1}(y)$, which is shown with gray arrows for two example points. In this example, we have used an exponential distribution. Hence, for $x \geq 0$, the probability density is $\varrho_X(x) = \lambda e^{-\lambda x}$ and the cumulated distribution function is $F(x) = 1 - e^{-\lambda x}$. Therefore, $x = F^{-1}(y) = -\frac{\ln(1-y)}{\lambda}$. We can see that using this method, many points end up close to 0 and only few points end up having high x-values - just as it is expected for an exponential distribution.

Note that the distribution does not change if we start with 1-y instead of y. For computational purposes, it therefore suffices to generate random numbers y in [0, 1] and then simply calculate

$$x = F^{-1}(y) = -\frac{1}{\lambda}\ln(y).$$

Proof of Correctness

Let F be a continuous cumulative distribution function, and let F^{-1} be its inverse function (using the infimum because CDFs are weakly monotonic and right-continuous):

$$F^{-1}(u) = \inf\{x \mid F(x) \geq u\} \qquad (0 < u < 1).$$

Claim: If U is a uniform random variable on (0, 1) then $F^{-1}(U)$ follows the distribution F.

Proof:

$$\begin{aligned} &\Pr(F^{-1}(U) \leq x) \\ &= \Pr(U \leq F(x)) \quad \text{(applying F, to both sides)} \\ &= F(x) \qquad\qquad\; \text{(because } \Pr(U \leq y) = y) \end{aligned}$$

Reduction of the Number of Inversions

In order to obtain a large number (lets say M) of samples one needs to perform the same number of inversions $F_X^{-1}(u)$ of the distribution F_X. One possible way to reduce the number of inversions to only a few while obtaining a large number of samples is the application of the so-called the Stochastic Collocation Monte Carlo sampler (SCMC sampler), within a polynomial chaos expansion framework, allows us the generation of any number of Monte Carlo samples based on only a few inversions of the original distribution and independent samples from a variable for which the inversions are analytically available, like for example the standard normal variable.

Inverse Sampling

In general, it is understood in the SRS methodology for qualitative characteristic that the attribute under study is not a rare attribute. If the attribute is rare, then the procedure of estimating the population proportion P by sample proportion n/N is not suitable. Some such situations are, e.g., estimation of frequency of rare type of genes, proportion of some rare type of cancer cells in a biopsy, proportion of rare type of blood cells affecting the red blood cells etc. In such cases, the methodology of inverse sampling can be used.

In the methodology of inverse sampling, the sampling is continued until a predetermined number of units possessing the attribute under study occur in the sampling which is useful for estimating the population proportion. The sampling units are drawn one-by-one with equal probability and without replacement. The sampling is discontinued as soon as the number of units in the sample possessing the characteristic or attribute equals a predetermined number.

Let m denotes the predetermined number indicating the number of units possessing the characteristic. The sampling is continued till m number of units are obtained. Therefore, the sample size n required to attain m becomes a random variable.

Probability Distribution Function of n

In order to find the probability distribution function of n, consider the stage of drawing of samples t such that at t = n, the sample size n completes the m units with attribute. Thus, the first (n - 1) draws would contain (m - 1) units in the sample possessing the characteristic out of NP units. Equivalently, there are (n - m) units which do not possess the characteristic out of NQ such units in the population. Note that the last draw must ensure that the units selected possess the characteristic.

So the probability distribution function of n can be expressed as.

$$P(n) = P\begin{pmatrix}\text{In a sample of } (n-1) \text{ units} \\ \text{drawn from N, } (m-1) \text{ units} \\ \text{will possess the attribute}\end{pmatrix} \times P\begin{pmatrix}\text{The unit drawn at} \\ \text{the } n^{th} \text{ draw will} \\ \text{possess the attribute}\end{pmatrix}$$

$$= \left[\frac{\binom{NP}{m-1}\binom{NQ}{n-m}}{\binom{N}{n-1}}\right]\left(\frac{NP-m+1}{N-n+1}\right), \quad n = m, m+1, \ldots\ldots, m+NQ.$$

Note that the first term (in square brackets) is derived using hypergeometric distribution as the probability of deriving a sample of size (n – 1) in which (m – 1) units are from NP units and (n – m) units are from NQ units. The second term $\frac{NP-m+1}{N-n+1}$ is the probability associated with last draw where it is assumed that we get the unit possessing the characteristic.

Note that $\sum_{n=m}^{m+NQ} P(n) = 1$

Estimate of Population Proportion

Consider the expectation of $\frac{m-1}{n-1}$

$$E\left(\frac{m-1}{n-1}\right) = \sum_{n=m}^{m+NQ}\left(\frac{m-1}{n-1}\right)P(n)$$

$$= \sum_{n=m}^{m+NQ}\left(\frac{m-1}{n-1}\right)\frac{\binom{NP}{m-1}\binom{NQ}{n-m}}{\binom{N}{n-1}} \cdot \frac{Np-m+1}{N-n+1}$$

$$= \sum_{n=m}^{m+NQ-1}\left(\frac{NP-m+1}{N-n+1}\right)\frac{\binom{NP-1}{m-2}\binom{NQ}{n-m}}{\binom{N-1}{n-2}}$$

which is obtained by replacing NP by NP – 1, m by (m – 1) and n by (n - 1) in the earlier step.

Thus

$$E\left(\frac{m-1}{n-1}\right) = P.$$

So

$$\hat{P} = \frac{m-1}{n-1}$$

is an unbiased estimator of P.

Estimate of variance of $\hat{P}$:

Now we derive an estimate of variance of $\hat{P}$. By definition

$$\begin{aligned} Var(\hat{P}) &= E\left(\hat{P}^2\right) - \left[E(\hat{P})\right]^2 \\ &= E\left(\hat{P}^2\right) - P^2 \end{aligned}$$

Thus

$$\widehat{Var}\left(\hat{P}\right) = \hat{P}^2 - \text{Estimate of } P^2$$

In order to obtain an estimate of P^2, consider the expectation of $\frac{(m-1)(m-2)}{(n-1)(n-2)}$, i.e.,

$$\begin{aligned} E\left[\frac{(m-1)(m-2)}{(n-1)(n-2)}\right] &= \sum_{n\geq m}\left[\frac{(m-1)(m-2)}{(n-1)(n-2)}\right]P(n) \\ &= \frac{P(NP-1)}{N-1}\sum_{n\geq m}\left(\frac{NP-m+1}{N-n+1}\right)\left[\frac{\binom{NP-2}{m-3}\binom{NQ}{n-m}}{\binom{N-2}{n-3}}\right] \end{aligned}$$

where the last term inside the square bracket is obtained by replacing NP by (NP-2), N by (n-2) and m by (m - 2) in the probability distribution function of hypergeometric distribution. This solves further to

$$E\left[\frac{(m-1)(m-2)}{(n-1)(n-2)}\right] = \frac{NP^2}{N-1} - \frac{P}{N-1}.$$

Thus an unbiased estimate of P^2 is

$$\begin{aligned} \text{Estimate of } P^2 &= \left(\frac{N-1}{N}\right)\frac{(m-1)(m-2)}{(n-1)(n-2)} + \frac{\hat{P}}{N} \\ &= \left(\frac{N-1}{N}\right)\frac{(m-1)(m-2)}{(n-1)(n-2)} + \frac{1}{N}\cdot\frac{m-1}{n-1} \end{aligned}$$

Finally, an estimate of variance of $\hat{P}$ is

$$\widehat{Var}(\hat{P}) = \hat{P}^2 - \text{Estimate of } P^2$$

$$= \left(\frac{m-1}{n-1}\right)^2 - \left[\frac{N-1}{N}.\frac{(m-1)(m-2)}{(n-1)(n-2)} + \frac{1}{N}\left(\frac{m-1}{n-1}\right)\right]$$

$$= \left(\frac{m-1}{n-1}\right)\left[\left(\frac{m-1}{n-1}\right) + \frac{1}{N}\left(1 - \frac{(N-1)(m-2)}{n-2}\right)\right]$$

For large *N*, the hypergeometric distribution tends to negative Binomial distribution with probability density function

$$\binom{n-1}{m-1} P^m Q^{n-m}$$

So

$$\hat{P} = \frac{m-1}{n-1}$$

and

$$\widehat{Var}(\hat{P}) = \frac{(m-1)(n-m)}{(n-1)^2(n-2)} = \frac{\hat{P}(1-\hat{P})}{n-2}$$

Estimation of Proportion for More than Two Classes

We have assumed up to now that there are only two classes in which the population can be divided based on a qualitative characteristic. There can be situations when the population is to be divided into more than two classes.

For example, the taste of a coffee can be divided into four categories very strong, strong, mild and very mild. Similarly in another example the damage to crop due to storm can be classified into categories like heavily damaged, damaged, minor damage and no damage etc.

These type of situations can be represented by dividing the population of size *N* into, say *k*, mutually exclusive classes $c_1 c_2, \ldots\ldots, c_k$ Corresponding to these classes, let

$P_1 = \frac{C_1}{N}, P_2 = \frac{C_2}{N}, \ldots\ldots, P_k = \frac{C_k}{n}$, be the proportions of units in the classes $c_1 c_2, \ldots\ldots, c_k$ respectively.

Let a sample of size *n* is observed such that $c_1 c_2, \ldots\ldots, c_k$ number of units have been drawn from $c_1 c_2, \ldots\ldots, c_k$ respectively. Then the probability of observing $c_1 c_2, \ldots\ldots, c_k$ is

$$P(c_1 c_2, \ldots\ldots, c_k) = \frac{\binom{C_1}{c_1}\binom{C_2}{c_2}\cdots\binom{C_k}{c_k}}{\binom{N}{n}}$$

The population proportions P_i can be estimated by $p_i = \frac{c_i}{n}, i = 1,2,\ldots,k.$

It can be easily shown that

$$E(p_i) = P_i, \quad i = 1,2,\ldots,k,$$
$$Var(p_i) = \frac{N-n}{N-1}\frac{P_iQ_i}{n} \text{ and } \widehat{Var}(p_i) = \frac{N-n}{N}\frac{p_iq_i}{n-1}$$

For estimating the number of units in the ith class,

$$\hat{C}_i = Np_i$$
$$Var(\hat{C}_i) = N^2 Var(p_i) \text{ and } \widehat{Var}(\hat{C}_i) = N^2 \widehat{Var}(p_i)$$

The confidence intervals can be obtained based on single p_i as in the case of two classes.

If *N* is large, then the probability of observing $c_1 c_2, \ldots, c_k$ can be approximated by multinomial distribution given by

$$P(c_1 c_2, \ldots, c_k) = \frac{n!}{c_1! c_2! \ldots c_k!} P_1^{c_1} P_1^{c_2} \ldots P_k^{c_k}$$

For this Distribution

$$E(p_i) = P_i, \quad i = 1,2,\ldots,k,$$
$$Var(p_i) = \frac{P_i(1-P_i)}{n} \text{ and } \widehat{Var}(\hat{p}_i) = \frac{p_i(1-p_i)}{n}$$

Rejection Sampling

In mathematics, rejection sampling is a basic technique used to generate observations from a distribution. It is also commonly called the acceptance-rejection method or "accept-reject algorithm" and is a type of Monte Carlo method. The method works for any distribution in $\mathbb{R}^m$ with a density.

Rejection sampling is based on the observation that to sample a random variable one can perform a uniformly random sampling of the 2D cartesian graph, and keep the samples in the region under the graph of its density function. Note that this property can be extended to N-dimension functions.

Description

To visualize the motivation behind rejection sampling, imagine graphing the density function of a random variable onto a large rectangular board and throwing darts at it. Assume that the darts are uniformly distributed around the board. Now remove all of the darts that are outside the area under the curve. The remaining darts will be distributed uniformly within the area under the curve, and the x-positions of these darts will be distributed according to the random variable's density. This is because there is the most room for the darts to land where the curve is highest and thus the probability density is greatest.

The visualization as just described is equivalent to a particular form of rejection sampling where the proposal distribution is uniform (hence its graph is a rectangle). The general form of rejection sampling assumes that the board is not necessarily rectangular but is shaped according to some distribution that we know how to sample from (for example, using inversion sampling), and which is at least as high at every point as the distribution we want to sample from, so that the former completely encloses the latter. Otherwise, there will be parts of the curved area we want to sample from that can never be reached. Rejection sampling works as follows:

1. Sample a point on the x-axis from the proposal distribution.
2. Draw a vertical line at this x-position, up to the curve of the proposal distribution.
3. Sample uniformly along this line from 0 to the maximum of the probability density function. If the sampled value is greater than the value of the desired distribution at this vertical line, return to step 1.

This algorithm can be used to sample from the area under any curve, regardless of whether the function integrates to 1. In fact, scaling a function by a constant has no effect on the sampled x-positions. Thus, the algorithm can be used to sample from a distribution whose normalizing constant is unknown, which is common in computational statistics.

Examples

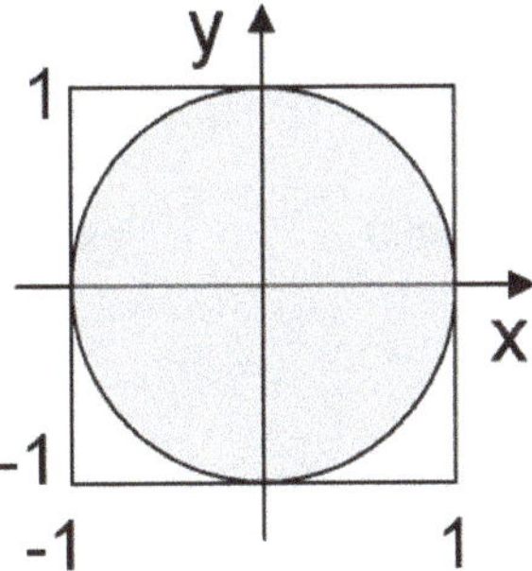

As a simple geometric example, suppose it is desired to generate a random point within the unit circle. Generate a candidate point (x, y) where x and y are independent uniformly distributed between -1 and 1. If it happens that $x^2 + y^2 \leq 1$ then the point is within the unit circle and should be accepted. If not then this point should be rejected and another candidate should be generated.

The ziggurat algorithm, a more advanced example, is used to efficiently generate normally-distributed pseudorandom numbers.

Theory

The rejection sampling method generates sampling values from a target distribution X with arbitrary probability density function $f(x)$ by using a proposal distribution Y with probability density $g(x)$. The idea is that one can generate a sample value from X by instead sampling from Y and accepting the sample from Y with probability $f(x)/(Mg(x))$, repeating the draws from Y until a value is accepted. M here is a constant, finite bound on the likelihood ratio $f(x)/g(x)$, satisfying $1 < M < \infty$ over the support of X ; in other words, M must satisfy $f(x) \leq Mg(x)$ for all values of x. Note that this requires that the support of Y must include the support of X —in other words, $g(x) > 0$ whenever $f(x) > 0$.

The validation of this method is the envelope principle: when simulating the pair $(x, v = u \cdot Mg(x))$, one produces a uniform simulation over the subgraph of $Mg(x)$. Accepting only pairs such that $u < f(x)/Mg(x)$ then produces pairs (x, v) uniformly distributed over the subgraph of $f(x)$ and thus, marginally, a simulation from $f(x)$.

This means that, with enough replicates, the algorithm generates a sample from the desired distribution $f(x)$. There are a number of extensions to this algorithm, such as the Metropolis algorithm and the combination with ratio-of-uniforms approach.

This method relates to the general field of Monte Carlo techniques, including Markov chain Monte Carlo algorithms that also use a proxy distribution to achieve simulation from the target distribution $f(x)$. It forms the basis for algorithms such as the Metropolis algorithm.

The unconditional acceptance probability is the proportion of proposed samples which are accepted, which is

$$\begin{aligned}\mathbb{P}\left(U \leq \frac{f(x)}{Mg(x)}\right) &= E\left[\mathbb{P}\left(U \leq \frac{f(x)}{Mg(x)} \mid x\right)\right] \\ &= E\left[\frac{f(x)}{Mg(x)}\right] \\ &= \int \frac{f(z)}{Mg(z)} g(z)dz \\ &= \frac{1}{M}\int f(z)dz \\ &= \frac{1}{M}\end{aligned}$$

where $U \sim \mathrm{Unif}(0,1)$, and the value of y each time is generated under the density function $g(x)$ of the proposal distribution Y.

The number of samples required from Y to obtain an accepted value thus follows a geometric distribution with probability $1/M$, which has mean M. Intuitively, M is the expected number of the iterations that are needed, as a measure of the computational complexity of the algorithm.

Rewrite the above equation,

$$M = \frac{1}{\mathbb{P}\left(U \le \frac{f(y)}{Mg(y)}\right)}$$

Note that $1 \le M < \infty$, due to the above formula, where $\mathbb{P}\left(U \le \frac{f(Y)}{Mg(Y)}\right)$ is a probability which can only take values in the interval $[0,1]$. When M is chosen closer to one, the unconditional acceptance probability is higher the less that ratio varies, since M is the upper bound for the likelihood ratio $f(x)/g(x)$. In practice, a value of M closer to 1 is preferred as it implies fewer rejected samples, on average, and thus fewer iterations of the algorithm. In this sense, one prefers to have M as small as possible (while still satisfying $f(x) \le Mg(x)$, which suggests that $g(x)$ should generally resemble $f(x)$ in some way. Note, however, that M cannot be equal to 1: such would imply that M, i.e. that the target and proposal distributions are actually the same distribution.

Rejection sampling is most often used in cases where the form of $f(x)$ makes sampling difficult. A single iteration of the rejection algorithm requires sampling from the proposal distribution, drawing from a uniform distribution, and evaluating the $f(x)/(Mg(x))$ expression. Rejection sampling is thus more efficient than some other method whenever M times the cost of these operations—which is the expected cost of obtaining a sample with rejection sampling—is lower than the cost of obtaining a sample using the other method.

Algorithm

The algorithm (used by John von Neumann and dating back to Buffon and his needle) to obtain a sample from distribution X with density f using samples from distribution Y with density g is as follows:

- Obtain a sample y from distribution Y and a sample u from $\mathrm{Unif}(0,1)$ (the uniform distribution over the unit interval).
- Check whether or not $u < f(y)/Mg(y)$.
    - If this holds, accept y as a sample drawn from f;
    - if not, reject the value of y and return to the sampling step.

The algorithm will take an average of M iterations to obtain a sample.

Advantages Over Sampling Using Naive Methods

Rejection sampling can be far more efficient compared with the Naive methods in some situations. For example, given a problem as sampling $X \sim F(\cdot)$ conditionally on X given the set A, i.e., $X \mid X \in A$, sometimes X can be easily simulated, using the Naive methods (e.g. by inverse transform sampling):

- Sample $X \sim F(\cdot)$ independently, and leave those satisfying $\{n \geq 1 : X_n \in A\}$
- Output: $\{X_1, X_2, \ldots, X_N : X_i \in A, i = 1, \ldots, N\}$

The problem is this sampling can be difficult and inefficient, if $\mathbb{P}(X \in A) \approx 0$. The expected number of iterations would be $\frac{1}{\mathbb{P}(X \in A)}$, which could be close to infinity. Moreover, even when you apply Rejection sampling method, it is always hard to optimize the bound M for the likelihood ratio. More often than not, M is large and the rejection rate is high, the algorithm can be very inefficient. The Natural Exponential Family (if it exists), also known as exponential tilting, provides a class of proposal distributions that can lower the computation complexity, the value of M and speed up the computations.

Examples: Working with Natural Exponential Families

Given a random variable $X \sim F(\cdot)$, $F(x) = \quad (X \leq x)$ is the target distribution. Assume for the simplicity, the density function can be explicitly written as $f(x)$. Choose the proposal as

$$\begin{aligned} F_\theta(x) &= \mathbb{E}\left[\exp(\theta X - \psi(\theta))\mathbb{I}(X \leq x)\right] \\ &= \int_{-\infty}^{x} e^{\theta y - \psi(\theta)} f(y)dy \\ g_\theta(x) = F_\theta'(x) &= e^{\theta x - \psi(\theta)} f(x) \end{aligned}$$

where $\psi(\theta) = \log(\mathbb{E}\exp(\theta X))$ and $\Theta = \{\theta : \psi(\theta) < \infty\}$. Clearly, $\{F_\theta(\cdot)\}_{\theta \in \Theta}$, is from a natural exponential family. Moreover, the likelihood ratio is

$$Z(x) = \frac{f(x)}{g_\theta(x)} = \frac{f(x)}{e^{\theta x - \psi(\theta)} f(x)} = e^{-\theta x + \psi(\theta)}$$

Note that $\psi(\theta) < \infty$ implies that it is indeed a log moment-generation function, that is, $\psi(\theta) = \log \mathbb{E}\exp(tX)\,|_{t=\theta} = \log M_X(t)\,|_{t=\theta}$. And it is easy to derive the log moment-generation function of the proposal and therefore the proposal's moments.

$$\begin{aligned} \psi_\theta(\eta) &= \log(\mathbb{E}_\theta \exp(\eta X)) = \psi(\theta + \eta) - \psi(\theta) < \infty \\ \mathbb{E}_\theta(X) &= \frac{\partial \psi_\theta(\eta)}{\partial \eta}\Big|_{\eta=0} \\ \mathrm{Var}_\theta(X) &= \frac{\partial^2 \psi_\theta(\eta)}{\partial^2 \eta}\Big|_{\eta=0} \end{aligned}$$

As a simple example, suppose under $F(\cdot)$, $X \sim N(\mu,\sigma^2)$, with $\psi(\theta) = \theta\mu + \frac{\sigma^2\theta^2}{2}$. The goal is to sample $X \mid X \in [b,\infty]$, $b > \mu$. The analysis goes as followed.

- Choose the form of the proposal distribution $F_\theta(\cdot)$, with log moment-generating function as $\psi_\theta(\eta) = \psi(\theta+\eta) - \psi(\eta) = \eta(\mu + \theta\sigma^2) + \frac{\sigma^2\eta^2}{2}$, which further implies it is a normal distribution $N(\mu + \theta\sigma^2, \sigma^2)$.
- Decide the well chosen θ^* for the proposal distribution. In this setup, the intuitive way to choose θ^* is to set $\mathbb{E}_\theta(X) = \mu + \theta\sigma^2 = b$, that is $\theta^* = \frac{b-\mu}{\sigma^2}$
- Explicitly write out the target, the proposal and the likelihood ratio

$$f_{X|X\geq b}(x) = \frac{f(x)\mathbb{I}(x \geq b)}{\mathbb{P}(x \geq b)}$$

$$g_{\theta^*}(x) = f(x)\exp(\theta^* x - \psi(\theta^*))$$

$$Z(x) = \frac{f_{X|X\geq b}(x)}{g_{\theta^*}(x)} = \frac{\exp(-\theta^* x + \psi(\theta^*))\mathbb{I}(x \geq b)}{\mathbb{P}(x \geq b)}$$

- Derive the bound M for the likelihood ratio $z(x)$, which is a decreasing function for $x \in [b,\infty]$, therefore

$$M = Z(b) = \frac{\exp(-\theta^* b + \psi(\theta^*))}{\mathbb{P}(X \geq b)} = \frac{\exp(-\frac{(b-\mu)^2}{2\sigma^2})}{\mathbb{P}(X \geq b)} = \frac{\exp(-\frac{(b-\mu)^2}{2\sigma^2})}{\mathbb{P}(N(0,1) \geq \frac{b-\mu}{\sigma})}$$

- Rejection sampling criterion: for $U \sim \text{Unif}(0,1)$, if

$$U \leq \frac{Z(x)}{M} = e^{-\theta^*(x-b)}\mathbb{I}(x \geq b)$$

holds, accept the value of X; if not, continue sampling new $X \sim_{\text{i.i.d.}} N(\mu + \theta^*\sigma^2, \sigma^2)$ and new $U \sim \text{Unif}(0,1)$ until acceptance.

For the above example, as the measurement of the efficiency, the expected number of the iterations the NEF-Based Rejection sampling method is of order b, that is $M(b) = O(b)$, while under the Naive method, the expected number of the iterations is $\frac{1}{\mathbb{P}(X \geq b)} = O(b \cdot e^{\frac{(b-\mu)^2}{2\sigma^2}})$, which is far more inefficient.

In general, exponential tilting, a parametric class of proposal distribution, solves the optimization problems conveniently, with its useful properties that directly characterize the distribution of the proposal. For this type of problem, to simulate X conditionally on $X \in A$, among the class of simple distributions, the trick is to use NEFs, which helps to gain some control over the complexity and considerably speed up the computation. Indeed, there are deep mathematical reasons for using NEFs.

Drawbacks

Rejection sampling can lead to a lot of unwanted samples being taken if the function being sampled is highly concentrated in a certain region, for example a function that has a spike at some location. For many distributions, this problem can be solved using an adaptive extension. In addition, as the dimensions of the problem get larger, the ratio of the embedded volume to the "corners" of the embedding volume tends towards zero, thus a lot of rejections can take place before a useful sample is generated, thus making the algorithm inefficient and impractical. In high dimensions, it is necessary to use a different approach, typically a Markov chain Monte Carlo method such as Metropolis sampling or Gibbs sampling. (However, Gibbs sampling, which breaks down a multi-dimensional sampling problem into a series of low-dimensional samples, may use rejection sampling as one of its steps.)

Adaptive Rejection Sampling

For many distributions, finding a proposal distribution that includes the given distribution without a lot of wasted space is difficult. An extension of rejection sampling that can be used to overcome this difficulty and efficiently sample from a wide variety of distributions (provided that they have log-concave density functions, which is in fact the case for most of the common distributions—even those whose *density* functions are not concave themselves!) is known as adaptive rejection sampling (ARS).

There are three basic ideas to this technique as ultimately introduced by Gilks in 1992:

1. If it helps, define your envelope distribution in log space (e.g. log-probability or log-density) instead. That is, work with $h(x) = \log g(x)$ instead of $g(x)$ directly.

    - Often, distributions that have algebraically messy density functions have reasonably simpler log density functions (i.e. when $f(x)$ is messy, $\log f(x)$ may be easier to work with or, at least, closer to piecewise linear).

2. Instead of a single uniform envelope density function, use a piecewise linear density function as your envelope instead.

    - Each time you have to reject a sample, you can use the value of $f(x)$ that you evaluated, to improve the piecewise approximation $h(x)$. This therefore reduces the chance that your next attempt will be rejected. Asymptotically, the probability of needing to reject your sample should converge to zero, and in practice, often very rapidly.

    - As proposed, any time we choose a point that is rejected, we tighten the envelope with another line segment that is tangent to the curve at the point with the same x-coordinate as the chosen point.

- A piecewise linear model of the proposal log distribution results in a set of piecewise exponential distributions (i.e. segments of one or more exponential distributions, attached end to end). Exponential distributions are well behaved and well understood. The logarithm of an exponential distribution is a straight line, and hence this method essentially involves enclosing the logarithm of the density in a series of line segments. This is the source of the log-concave restriction: if a distribution is log-concave, then its logarithm is concave (shaped like an upside-down U), meaning that a line segment tangent to the curve will always pass over the curve.
- If not working in log space, a piecewise linear density function can also be sampled via triangle distributions

3. We can take even further advantage of the (log) concavity requirement, to potentially avoid the cost of evaluating $f(x)$ when your sample *is* accepted.
    - Just like we can construct a piecewise linear upper bound (the "envelope" function) using the values of $h(x)$ that we had to evaluate in the current chain of rejections, we can also construct a piecewise linear lower bound (the "squeezing" function) using these values as well.
    - Before evaluating (the potentially expensive) $f(x)$ to see if your sample will be accepted, we may *already know* if it will be accepted by comparing against the (ideally cheaper) $g_l(x)$ (or $h_l(x)$ in this case) squeezing function that have available.
    - This squeezing step is optional, even when suggested by Gilks. At best it saves you from only one extra evaluation of your (messy and/or expensive) target density. However, presumably for particularly expensive density functions (and assuming the rapid convergence of the rejection rate toward zero) this can make a sizable difference in ultimate runtime.

The method essentially involves successively determining an envelope of straight-line segments that approximates the logarithm better and better while still remaining above the curve, starting with a fixed number of segments (possibly just a single tangent line). Sampling from a truncated exponential random variable is straightfoward. Just take the log of a uniform random variable (with appropriate interva and corresponding truncation).

Unfortunately, ARS can only be applied from sampling from log-concave target densities. For this reason, several extensions of ARS have been proposed in literature for tackling non-log-concave target distributions. Furthermore, different combinations of ARS and the Metropolis-Hastings method have been designed in order to obtain a universal sampler that builds a self-tuning proposal densities (i.e., a proposal automatically constructed and adapted to the target). This class of methods are often called as Adaptive Rejection Metropolis Sampling (ARMS) algorithms.

The resulting adaptive techniques can be always applied but the generated samples are correlated in this case (although the correlation vanishes quickly to zero as the number of iterations grows).

Nonprobability Sampling

Sampling is the use of a subset of the population to represent the whole population or to inform about (social) processes that are meaningful beyond the particular cases, individuals or sites studied. Probability sampling, or random sampling, is a sampling technique in which the probability of getting any particular sample may be calculated. Nonprobability sampling does not meet this criterion and, as any methodological decision, should adjust to the research question that one envisages to answer. Nonprobability sampling techniques *are not intended to* be used to infer from the sample to the general population in statistical terms. Instead, for example, grounded theory can be produced through iterative non-probability sampling until theoretical saturation is reached (Strauss and Corbin, 1990).

Thus, one cannot say the same on the basis of a nonprobability sample than on the basis of a probability sample. The grounds for drawing generalizations (e.g., propose new theory, propose policy) from studies based on nonprobability samples are based on the notion of "theoretical saturation" and "analytical generalization" (Yin, 2014) instead of on statistical generalization. Researchers working with the notion of purposive sampling assert that while probability methods are suitable for large-scale studies concerned with representativeness, non-probability approaches are more suitable for in-depth qualitative research in which the focus is often to understand complex social phenomena (e.g., Marshall 1996; Small 2009). One of the advantages of nonprobability sampling is its lower cost compared to probability sampling. Moreover, the in-depth analysis of a small-N purposive sample or a case study enables the "discovery" and identification of patterns and causal mechanisms that do not draw time and context-free assumptions.

Non-probability sampling is often not appropriate in statistical quantitative research, though, as these assertions raise some questions —how can one understand a complex social phenomenon by drawing only the most convenient expressions of that phenomenon into consideration? What assumption about homogeneity in the world must one make to justify such assertions? Alas, the consideration that research can only be based in statistical inference focuses on the problems of bias linked to nonprobability sampling and acknowledges only one situation in which a non-probability sample can be appropriate —if one is interested *only* in the specific cases studied (for example, if one is interested in the Battle of Gettysburg), one does not need to draw a probability sample from similar cases (Lucas 2014a).

Non-probability sampling is however widely used in qualitative research. Examples of nonprobability sampling include:

- Convenience, haphazard or accidental sampling - members of the population are chosen based on their relative ease of access. To sample friends, co-workers, or shoppers at a single mall, are all examples of convenience sampling. Such samples are biased because researchers may unconsciously approach some kinds of respondents and avoid others (Lucas 2014a), and respondents who volunteer for a study may differ in unknown but important ways from others (Wiederman 1999).

- Snowball sampling - The first respondent refers an acquaintance. The friend also refers a friend, and so on. Such samples are biased because they give people with more social connections an unknown but higher chance of selection (Berg 2006), but lead to higher response rates.

- Judgmental sampling or purposive sampling - The researcher chooses the sample based on who they think would be appropriate for the study. This is used primarily when there is a limited number of people that have expertise in the area being researched, or when the interest of the research is on a specific field or a small group. Different types of purposive sampling include:

- Deviant case - The researcher obtains cases that substantially differ from the dominant pattern (a special type of purposive sample). The case is selected in order to obtain information on unusual cases that can be specially problematic or specially good.

- Case study - The research is limited to one group, often with a similar characteristic or of small size.

- Ad hoc quotas - A quota is established (e.g. 65% women) and researchers are free to choose any respondent they wish as long as the quota is met.

Nonprobability sampling should not intend to meet the same type of results neither to be assessed with the quality criteria of probabilistic sampling (Steinke, 2004).

Studies intended to use probability sampling sometimes end up using nonprobability samples because of characteristics of the sampling method. For example, using a sample of people in the paid labor force to analyze the effect of education on earnings is to use a non-probability sample of persons who could be in the paid labor force. Because the education people obtain could determine their likelihood of being in the paid labor force, technically the sample in the paid labor force is a nonprobability sample for the question at issue. In such cases results are biased.

The statistical model one uses can also render the data a non-probability sample. For example, Lucas (2014b) notes that several published studies that use multilevel modeling have been based on samples that are probability samples in general, but nonprobability samples for one or more of the levels of analysis in the study. Evidence indicates that in such cases the bias is poorly behaved, such that inferences from such analyses are unjustified.

These problems occur in the academic literature, but they may be more common in non-academic research. For example, in public opinion polling by private companies (or other organizations unable to require response), the sample can be self-selected rather than random. This often introduces an important type of error: self-selection bias. This error sometimes makes it unlikely that the sample will accurately represent the broader population. More important, this error makes it impossible to establish that the sample represents the broader population. Volunteering for the sample may be determined by characteristics such as submissiveness or availability. The samples in such surveys should be treated as non-probability samples of the population, and the validity of the findings based on them is unknown and cannot be established.

References

- Park, Sung Y.; Bera, Anil K. (2009). "Maximum entropy autoregressive conditional heteroskedasticity model". Journal of Econometrics. Elsevier. 150 (2): 219–230. doi:10.1016/j.jeconom.2008.12.014
- Vitter, Jeffrey S. (1985-03-01). "Random Sampling with a Reservoir". ACM Trans. Math. Softw. 11 (1): 37–57. ISSN 0098-3500. doi:10.1145/3147.3165
- Evans, M.; Swartz, T. (1998-12-01). "Random Variable Generation Using Concavity Properties of Transformed Densities". Journal of Computational and Graphical Statistics. 7 (4): 514–528. JSTOR 1390680. doi:10.2307/1390680
- Sunter, A. B. (1977-01-01). "List Sequential Sampling with Equal or Unequal Probabilities without Replacement". Applied Statistics. 26 (3). JSTOR 10.2307/2346966. doi:10.2307/2346966
- Yates, Daniel S.; David S. Moore; Daren S. Starnes (2008). The Practice of Statistics, 3rd Ed. Freeman. ISBN 978-0-7167-7309-2
- Görür, Dilan; Teh, Yee Whye (2011-01-01). "Concave-Convex Adaptive Rejection Sampling". Journal of Computational and Graphical Statistics. 20 (3): 670–691. ISSN 1061-8600. doi:10.1198/jcgs.2011.09058
- Bishop, Christopher (2006). "11.4: Slice sampling". Pattern Recognition and Machine Learning. Springer. ISBN 0-387-31073-8
- Neal, Radford M. (2003). "Slice Sampling". Annals of Statistics. 31 (3): 705–767. MR 1994729. Zbl 1051.65007. doi:10.1214/aos/1056562461
- Wiederman, Michael W. (1999). "Volunteer bias in sexuality research using college student participants." Journal of Sex Research, 36: 59-66, doi:10.1080/00224499909551968
- Small, Mario L. (2009). "'How many cases do I need?' On science and the logic of case selection in field-based research." Ethnography 10: 5–38. doi:10.1177/1466138108099586
- Fan, C. T.; Muller, Mervin E.; Rezucha, Ivan (1962-06-01). "Development of Sampling Plans by Using Sequential (Item by Item) Selection Techniques and Digital Computers". Journal of the American Statistical Association. 57 (298): 387–402. ISSN 0162-1459. doi:10.1080/01621459.1962.10480667

3

Stratified Sampling and Jackknife Resampling

Stratified sampling is the method of sampling that involves the division of the population. It categorizes the population into smaller groups which are known as strata. The groups are formed on the basis of similarity of attributes or characteristics. Stratified sampling is best understood in confluence with the major topics listed in the following chapter.

Stratified Sampling

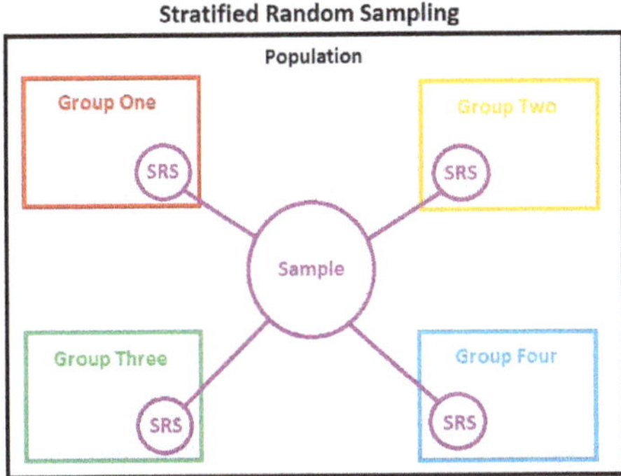

Stratified Random Sampling

In statistics, stratified sampling is a method of sampling from a population.

In statistical surveys, when subpopulations within an overall population vary, it is advantageous to sample each subpopulation (stratum) independently. Stratification is the process of dividing members of the population into homogeneous subgroups before sampling. The strata should be mutually exclusive: every element in the population must be assigned to only one stratum. The strata should also be collectively exhaustive: no population element can be excluded. Then simple random sampling or systematic sampling is applied within each stratum. This often improves the representativeness of the sample by reducing sampling error. It can produce a weighted mean that has less variability than the arithmetic mean of a simple random sample of the population.

In computational statistics, stratified sampling is a method of variance reduction when Monte Carlo methods are used to estimate population statistics from a known population.

Example

Assume that we need to estimate average number of votes for each candidate in an election. Assume that country has 3 towns: Town A has 1 million factory workers, Town B has 2 million office workers and Town C has 3 million retirees. We can choose to get a random sample of size 60 over entire population but there is some chance that the random sample turns out to be not well balanced across these towns and hence is biased causing a significant error in estimation. Instead if we choose to take a random sample of 10, 20 and 30 from Town A, B and C respectively then we can produce a smaller error in estimation for the same total size of sample.

Stratified Sampling Strategies

1. *Proportionate allocation* uses a sampling fraction in each of the strata that is proportional to that of the total population. For instance, if the population consists of X total individuals, m of which are male and f female (and where $m + f = X$), then the relative size of the two samples ($x1 = m/X$ males, $x2 = f/X$ females) should reflect this proportion.

2. *Optimum allocation* (or *disproportionate allocation*) - Each stratum is proportionate to the standard deviation of the distribution of the variable. Larger samples are taken in the strata with the greatest variability to generate the least possible sampling variance.

Stratified sampling ensures that at least one observation is picked from each of the strata, even if probability of it being selected is close to 0. Hence the statistical properties of the population may not be preserved if there are thin strata. A rule of thumb that is used to ensure this is that the population should consist of no more than six strata, but depending on special cases the rule can change - for example if there are 100 strata each with 1 million observations, it is perfectly fine to do a 10% stratified sampling on them.

A real-world example of using stratified sampling would be for a political survey. If the respondents needed to reflect the diversity of the population, the researcher would specifically seek to include participants of various minority groups such as race or religion, based on their proportionality to the total population as mentioned above. A stratified survey could thus claim to be more representative of the population than a survey of simple random sampling or systematic sampling.

Advantages

The reasons to use stratified sampling rather than simple random sampling include

1. If measurements within strata have lower standard deviation, stratification gives smaller error in estimation.

2. For many applications, measurements become more manageable and/or cheaper when the population is grouped into strata.

3. It is often desirable to have estimates of population parameters for groups within the population.

If the population density varies greatly within a region, stratified sampling will ensure that estimates can be made with equal accuracy in different parts of the region, and that comparisons of sub-regions can be made with equal statistical power. For example, in Ontario a survey taken throughout the province might use a larger sampling fraction in the less populated north, since the disparity in population between north and south is so great that a sampling fraction based on the provincial sample as a whole might result in the collection of only a handful of data from the north.

Randomized stratification can also be used to improve population representativeness in a study.

Disadvantages

Stratified sampling is not useful when the population cannot be exhaustively partitioned into disjoint subgroups. It would be a misapplication of the technique to make subgroups' sample sizes proportional to the amount of data available from the subgroups, rather than scaling sample sizes to subgroup sizes (or to their variances, if known to vary significantly e.g. by means of an F Test). Data representing each subgroup are taken to be of equal importance if suspected variation among them warrants stratified sampling. If subgroup variances differ significantly and the data needs to be stratified by variance, it is not possible to simultaneously make each subgroup sample size proportional to subgroup size within the total population. For an efficient way to partition sampling resources among groups that vary in their means, variance and costs. The problem of stratified sampling in the case of unknown class priors (ratio of subpopulations in the entire population) can have deleterious effect on the performance of any analysis on the dataset, e.g. classification. In that regard, minimax sampling ratio can be used to make the dataset robust with respect to uncertainty in the underlying data generating process.

Mean and Variance

The mean and variance of stratified random sampling is given by,

$$\mu_s = \frac{1}{N}\sum_{h=1}^{L} N_h \mu_h$$

$$\sigma_s^2 = \sum_{h=1}^{L}\left(\frac{N_h}{N}\right)^2\left(\frac{N_h - n_h}{N_h}\right)\frac{\sigma_h^2}{n_h}$$

where,

N = Size of entire population, should equal to sum of all stratum sizes

N_h = Size of stratum

n_h = Number of observations in stratum

L = Count of strata

σ_h = sample standard deviation of stratum

μ_h = sample mean of stratum.

Strata Size Calculation

For proportional allocation strategy, the size of the sample in each stratum is taken in proportion to the size of the stratum. Suppose that in a company there are the following staff:

- male, full-time: 90
- male, part-time: 18
- female, full-time: 9
- female, part-time: 63
- total: 180

and we are asked to take a sample of 40 staff, stratified according to the above categories.

The first step is to calculate the percentage of each group of the total.

- % male, full-time = 90 ÷ 180 = 50%
- % male, part-time = 18 ÷ 180 = 10%
- % female, full-time = 9 ÷ 180 = 5%
- % female, part-time = 63 ÷ 180 = 35%

This tells us that of our sample of 40,

- 50% (20 individuals) should be male, full-time.
- 10% (4 individuals) should be male, part-time.
- 5% (2 individuals) should be female, full-time.
- 35% (14 individuals) should be female, part-time.

Another easy way without having to calculate the percentage is to multiply each group size by the sample size and divide by the total population size (size of entire staff):

- male, full-time = 90 × (40 ÷ 180) = 20
- male, part-time = 18 × (40 ÷ 180) = 4
- female, full-time = 9 × (40 ÷ 180) = 2
- female, part-time = 63 × (40 ÷ 180) = 14

An important objective in any estimation problem is to obtain an estimator of a population parameter which can take care of all salient features of the population. If the population is homogeneous with respect to the characteristic under study, then the method of simple random sampling will yield a homogeneous sample and in turn, the sample mean will serve as a good estimator of population mean. Thus, if the population is homogeneous with respect to the characteristic under study, then the sample drawn through simple random sampling is expected to provide a representative sample.

Moreover, the variance of sample mean not only depends on the sample size and sampling fraction but also on the population variance. In order to increase the precision of an estimator, we have to use a sampling scheme which reduces the heterogeneity in the population. If the population is heterogeneous with respect to the characteristic under study, then one such sampling procedure is stratified sampling.

Example

In order to find the average height of students in a school of class 1 to class 12, the height varies a lot as the students in class 1 are of age around 6 years and students in class 10 are of age around 16 years. So one can divide all the students into different subpopulations or strata such as:

Students of class 1, 2 and 3: Stratum 1

Students of class 4, 5 and 6: Stratum 2

Students of class 7, 8 and 9: Stratum 3

Students of class 10, 11 and 12: Stratum 4

Now draw the samples by SRS from each of the strata 1, 2 ,3 and 4. All the drawn samples combined together will constitute the final stratified sample for further analysis.

Notations

We use the following symbols and notations:

N : Population size

K : Number of strata

N_i : Number of sampling units in i^{th} strata

$N = \sum_{i=1}^{k} N_i$

n_i : Numbers of sampling units to be drawn from i^{th} stratum

$n = \sum_{i=1}^{k} n_i$: Total sample size.

Procedure of Stratified Sampling

- Divide the population of N units into k strata. Let the ith stratum have $N_1, i = 1, 2, ..., k$ number of units.
- Strata are constructed such that they are non-overlapping and homogeneous with respect to the characteristic under study such that

$$\sum_{i=1}^{k} N_i = N.$$

- Draw a sample of size n_i from i^{th} $(i = 1, 2, ..., k)$ stratum using SRS (preferably WOR) independently from each stratum.
- All the sampling units drawn from each stratum will constitute a stratified sample of size $n = \sum_{i=1}^{k} n_i$.

Difference Between Stratified and Cluster Sampling Schemes

In stratified sampling, the strata are constructed such that they are

- within homogeneous and
- among heterogeneous.

In cluster sampling, the clusters are constructed such that they are

- within heterogeneous and
- among homogeneous.

Issue in Estimation in Stratified Sampling

Divide the population of N units into k strata. Let the ith stratum have $N_1, i = 1, 2, ..., k$ number of units. Note that there are k independent samples drawn through SRS of sizes $n_1, n_2, ..., n_k$. So, one can have kestimators of a parameter based on sizes Our interest is not to have k different estimators of the parameters but ultimate goal is to have a single estimator. In this case, an important issue is how to combine the different sample information together into one estimator which is good enough to provide the information about the parameter.

We now consider the estimation of population mean and population variance from a stratified sample.

Estimation of Population Mean and its Variance

Let

Y :characteristic under study,

y_{ij} :value of jth unit in i^{th} stratum $j = 1,2,...,n_i$ $i = 1,2,....,k,$

$\bar{Y}_i = \frac{1}{N_i}\sum_{j=1}^{N_i} y_{ij}$: population mean of ith stratum

$\bar{y}_i = \frac{1}{n_i}\sum_{j=1}^{n_i} y_{ij}$: sample mean of units from i^{th} stratum $j = 1,2,...,n$, $i = 1,2,....,k,$

$\bar{Y} = \frac{1}{N}\sum_{i=1}^{k} N_i \bar{Y}_i = \sum_{i=1}^{k} w_i \bar{Y}_i$: population mean where $w_i = \frac{N_i}{N}$

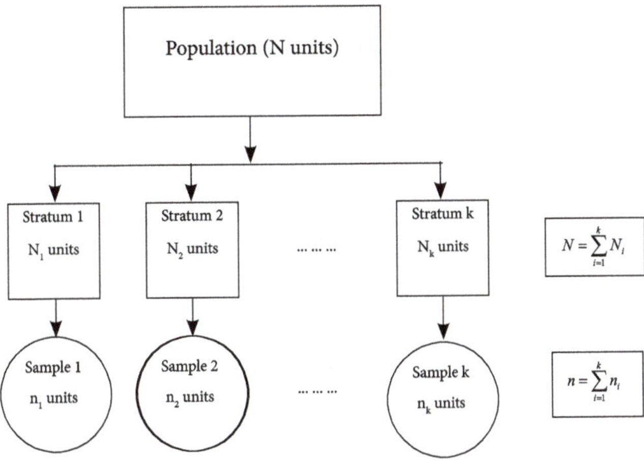

Estimation of Population Mean

First we discuss the estimation of population mean.

Note that the population mean is defined as the weighted arithmetic mean of stratum means in case of stratified sampling where the weights are provided in terms of strata sizes.

Based on the expression $\bar{Y} = \frac{1}{N}\sum_{i=1}^{k} N_i \bar{Y}_i$ one may choose the sample mean

$$\bar{y} = \frac{1}{n}\sum_{i=1}^{k} n_i \bar{y}_i$$

as a possible estimator of $\bar{Y}$

Since the sample in each stratum is drawn by SRS, so

$$E(\bar{y}_i) = \bar{Y}_i,$$

thus

$$\begin{aligned} E(\bar{y}) &= \frac{1}{n}\sum_{i=1}^{k} n_i E(\bar{y}_i) \\ &= \frac{1}{n}\sum_{i=1}^{k} n_i \bar{Y}_i \\ &\neq \frac{1}{n}\sum_{i=1}^{k} n_i \bar{Y}_i \\ &\neq \bar{Y} \end{aligned}$$

and $\bar{y}$ turns out to be a biased estimator of $\bar{Y}$. Based on this, one can modify $\bar{y}$ so as to obtain an unbiased estimator of $\bar{Y}$. Consider the stratum mean which is defined as the weighted arithmetic mean of strata sample means with strata sizes as weights given by

$$\bar{y}_{st} = \frac{1}{N}\sum_{i=1}^{k} N_i \bar{y}_i.$$

Now

$$\begin{aligned} E(\bar{y}_{st}) &= \frac{1}{N}\sum_{i=1}^{k} N_i E(\bar{y}_i) \\ &= \frac{1}{N}\sum_{i=1}^{k} N_i \bar{Y}_i \\ &= \bar{Y} \end{aligned}$$

Thus $\bar{y}_{st}$ is an unbiased estimator of $\bar{Y}$.

Variance of $\bar{y}_{st}$:

$$\operatorname{Var}(\bar{y}_{st}) = \sum_{i=1}^{k} w_i^2 \operatorname{Var}(\bar{y}_i) + \sum_{i(\neq j)=1}^{k} \sum_{j=1}^{n_i} w_i w_j \operatorname{Cov}(\bar{y}_i, \bar{y}_j)$$

Since all the samples have been drawn independently from each strata by SRSWOR so

$$\operatorname{Cov}(\bar{y}_i, \bar{y}_j) = 0$$

$$\operatorname{Var}(\bar{y}_i) = \frac{N_i - n_i}{N_i n_i} S_i^2$$

Where

$$S_i^2 = \frac{1}{N_i - 1}\sum_{j=1}^{N_i} \left(Y_{ij} - \bar{Y}_i\right)^2.$$

Thus

$$Var(\bar{y}_{st}) = \sum_{i=1}^{k} w_i^2 \frac{N_i - n_i}{N_i n_i} S_i^2$$

$$= \sum_{i=1}^{k} w_i^2 \left(1 - \frac{n_i}{N_i}\right) \frac{S_i^2}{n_i}.$$

Observe that $Var(\bar{y}_{st})$ is small when S_i^2 is small. This observation suggests how to construct the strata . If S_i^2 is small for all i = 1,2,...,k, then $Var(\bar{y}_{st})$ will also be small. That is why it was mentioned earlier that the strata are to be constructed such that they are within homogeneous, i.e., S_i^2 is small and among heterogeneous.

For example, the units in geographical proximity will tend to be closer. The consumption pattern in households will be similar within a lower income group housing society and within a higher income group housing society whereas they will differ a lot between the two housing societies based on income.

Estimate of variance

Since the samples have been drawn by SRSWOR, so

$$E(s_i^2) = S_i^2$$

Where $s_i^2 = \frac{1}{n_i - 1} \sum_{j=1}^{n_i} (y_{ij} - \bar{y}_i)^2$

and $\widehat{Var}(\bar{y}_i) = \frac{N_i - n_i}{N_i n_i} s_i^2$

so $\widehat{Var}(\bar{y}_{st}) = \sum_{j=1}^{k} w_i^2 \widehat{Var}(\bar{y}_i)$

$$= \sum_{i=1}^{k} w_i^2 \left(\frac{N_i - n_i}{N_i n_i} s_i^2 \right)$$

Note: If SRSWR is used instead of SRSWOR for drawing the samples from each stratum, then in this case

$$\bar{y}_{st} = \sum_{i=1}^{k} w_i \bar{y}_i$$

$$E(\bar{y}_{st}) = \bar{Y}$$

$$Var(\bar{y}_{st}) = \sum_{i=1}^{k} w_i^2 \left(\frac{N_i - 1}{N_i n_i} \right) S_i^2 = \sum_{i=1}^{k} w_i^2 \frac{\sigma_i^2}{n_i}$$

$$\widehat{Var}(\bar{y}_{st}) = \sum_{i=1}^{k} \frac{w_i^2 s_i^2}{n_i}$$

where $\sigma_i^2 = \frac{1}{N_i}\sum_{j=1}^{N_i}\left(y_{ij} - \bar{y}_i\right)^2$.

Advantages of Stratified Sampling

1. Data of known precision may be required for certain parts of the population. This can be accomplished with a more careful investigation to few strata.

Example: In order to know the direct impact of hike in petrol prices, the population can be divided into strata like lower income group, middle income group and higher income group. Obviously, the higher income group is more affected than the lower income group. So more careful investigation can be made in the higher income group strata.

2. Sampling problems may differ in different parts of the population.

Example: To study the consumption pattern of households, the people living in houses, hotels, hospitals, prison etc. are to be treated differently.

3. Administrative convenience can be exercised in stratified sampling.

Example: In taking a sample of villages from a big state, it is more administratively convenient to consider the districts as strata so that the administrative setup at district level may be used for this purpose. Such administrative convenience and convenience in organization of field work are important aspects in national level surveys.

4. Full cross-section of population can be obtained through stratified sampling. It may be possible in SRS that some large part of the population may remain unrepresented. Stratified sampling enables one to draw a sample representing different segments of the population to any desired extent. The desired degree of representation of some specified parts of population is also possible.

5. Substantial gain in efficiency is achieved if strata are formed intelligently.

6. In case of skewed population, use of stratification is of importance since larger weight may have to be given for the few extremely large units which in turn reduces the sampling variability.

7. When estimates are required not only for the population but also for the subpopulations, then stratified sampling is helpful.

8. When the sampling frame for subpopulations is more easily available than the sampling frame for whole population.

9. If population is large, then it is convenient to sample separately from the strata rather than the entire population.

10. The population mean or population total can be estimated with higher precision by suitably providing the weights to the estimates obtained from each stratum.

Allocation Problem and Choice of Sample Sizes in Different Strata

1. Equal Allocation

Choose the sample size n_i to be same for all the strata. Draw samples of equal size from each strata. Let n be the sample size and k be the number of strata, then $n_i = \frac{n}{k}$ for all $i = 1, 2, \ldots, k$.

2. Proportional Allocation

For fixed k, select ni such that it is proportional to stratum size N_i, i.e.,

$$n_i \propto N_i$$
$$\text{or } n_i = CN_i$$

where C is the constant of proportionality.

$$\sum_{i=1}^{k} n_i = \sum_{i=1}^{k} CN_i$$

$$\text{or } n = CN$$

$$\Rightarrow C = \frac{n}{N}.$$

Thus $n_i = \left(\frac{n}{N}\right) N_i$.

Such allocation arises from the considerations like operational convenience.

Neyman or Optimum Allocation

This allocation considers the size of strata as well as variability

$$n_i \propto N_i S_i$$
$$n_i = C^* N_i S_i$$

where C* is the constant of proportionality

$$\sum_{i=1}^{k} n_i = \sum_{i=1}^{k} C^* N_i S_i$$

or $n = C^* \sum_{i=1}^{k} N_i S_i$

or $C^* = \frac{n}{\sum_{i=1}^{k} N_i S_i}$.

Thus $n_i = \frac{n N_i S_i}{\sum_{i=1}^{k} N_i S_i}$.

This allocation arises when the $Var(\bar{y}_{st})$ is minimized subject to the constraint $\sum_{i=1}^{k} n_i$ (pre-specified).

There are some limitations of optimum allocation. The knowledge of $S_i (i = 1,2,....,k)$ is needed to know n_i.

If there are more than one characteristics, then they may lead to conflicting allocation.

Choice of Sample Size Based on Cost of Survey and Variability

The cost of survey depends upon the nature of survey. A simple choice of cost function is

$$C = C_0 + \sum_{i=1}^{k} C_i n_i$$

where

C : total cost

C_0 : overhead cost, e.g., setting up of office, training people etc.

C_i : cost per unit in the ith stratum

$\sum_{i=1}^{k} C_i n_i$: total cost within sample.

To find n_i under this cost function, consider the Lagrangian function with Lagrangian multiplier λ as

$$\begin{aligned}\phi &= Var(\bar{y}_{st}) + \lambda^2 (C - C_0)\\ &= \sum_{i=1}^{k} w_i^2 \left(\frac{1}{n_i} - \frac{1}{N_i}\right) S_i^2 + \lambda^2 \sum_{i=1}^{k} C_i n_i\\ &= \sum_{i=1}^{k} \frac{w_i^2 S_i^2}{n_i} + \lambda^2 \sum_{i=1}^{k} C_i n_i - \sum_{i=1}^{k} \frac{w_i^2 S_i^2}{N_i}\\ &= \sum_{i=1}^{k} \left[\frac{w_i S_i}{\sqrt{n_i}} - \lambda \sqrt{C_i n_i}\right]^2 + \text{terms independent of } n_i\end{aligned}$$

Thus ϕ is minimum when

$$\frac{w_i S_i}{\sqrt{n_i}} = \lambda \sqrt{C_i n_i} \text{ for all i}$$

or $n_i = \frac{1}{\lambda} \frac{w_i S_i}{\sqrt{C_i}}$.

How to determine λ?

There are two ways to determine λ

i. Minimize variability for fixed cost

ii. Minimize cost for given variability

Minimize Variability for Fixed Cost

Let $C = C_0^*$ be fixed.

So $\sum_{i=1}^{k} C_i n_i = C_0^*$

or $\sum_{i=1}^{k} C_i \frac{w_i S_i}{\lambda\sqrt{C_i}}. = C_0^*$

or $\lambda = \frac{\sum_{i=1}^{k} \sqrt{C_i} w_i S_i}{C_0^*}.$

Substituting λ in the expression for $n_i = \frac{1}{\lambda}\frac{w_i S_i}{\sqrt{C_i}}$, the optimum n_i is obtained as

$$n_i^* = \frac{w_i S_i}{\sqrt{C_i}}\left(\frac{C_0^*}{\sum_{i=1}^{k} \sqrt{C_i} w_i S_i}\right).$$

The required sample size to estimate $\bar{Y}$ such that the variance is minimum for given cost $C = C_0^*$ is

$$n = \sum_{i=1}^{k} n_i^*$$

Minimize Cost for Given Variability

Let $V = V_0$ be prespecified variance. Now determine n_i such that

$$\sum_{i=1}^{k}\left(\frac{1}{n_i} - \frac{1}{N_i}\right) w_i^2 S_i^2 = V_0$$

$$\text{or} \quad \sum_{i=1}^{k} \frac{w_i^2 S_i^2}{n_i} = V_0 + \sum_{i=1}^{k} \frac{w_i^2 S_i^2}{N_i}$$

$$\text{or} \quad \sum_{i=1}^{k} \frac{\lambda\sqrt{C_i}}{w_i S_i} w_i^2 S_i^2 = V_0 + \sum_{i=1}^{k} \frac{w_i^2 S_i^2}{N_i}$$

$$\text{or} \quad \lambda = \frac{V_0 + \sum_{i=1}^{k} \frac{w_i^2 S_i^2}{N_i}}{\sum_{i=1}^{k} w_i S_i \sqrt{C_i}} \left(\text{after substituting } n_i = \frac{1}{\lambda}\frac{w_i S_i}{\sqrt{C_i}}\right)$$

Thus the optimum n_i is

$$\tilde{n}_i = \frac{w_i S_i}{\sqrt{C_i}} \left(\frac{\sum_{i=1}^{k} w_i S_i \sqrt{C_i}}{V_0 + \sum_{i=1}^{k} \frac{w_i^2 S_i^2}{N_i}} \right)$$

So the required sample size to estimate $\bar{Y}$ such that cost C is minimum for a prespecified variance V_0 is $n = \sum_{i=1}^{k} \tilde{n}_i$.

Sample size under proportional allocation for fixed cost and for fixed variance

(i) If cost $C = C_0$ is fixed then $C_0 = \sum_{i=1}^{k} C_i n_i$.

Under proportional allocation, $n_i = \frac{n}{N} N_i = n w_i$.

So $C_0 = n \sum_{i=1}^{k} w_i C_i$

or $n = \frac{C_0}{\sum_{i=1}^{k} w_i C_i}$

Thus $n_i = \frac{C_0 w_i}{\sum w_i C_i}$

The required sample size to estimate $\bar{Y}$ in this case is $n = \sum_{i=1}^{k} n_i$

Variances Under Different Allocation

Now we derive the variance of $\bar{y}_{st}$ under proportional and optimum allocations.

(i) Proportional allocation

Under proportional allocation,

$$n_i = \frac{n}{N} N_i$$

and

$$Var(\bar{y})_{st} = \sum_{i=1}^{k} \left(\frac{N_i - n_i}{N_i n_i} \right) w_i^2 S_i^2$$

$$Var_{prop}(\bar{y})_{st} = \sum_{i=1}^{k} \left(\frac{N_i - \frac{n}{N} N_i}{N_i \frac{n}{N} N_i} \right) \left(\frac{N_i}{N} \right)^2 S_i^2$$

$$= \frac{N-n}{Nn} \sum_{i=1}^{k} \frac{N_i S_i^2}{N}$$

$$= \frac{N-n}{Nn} \sum_{i=1}^{k} w_i S_i^2 .$$

(ii) Optimum allocation

Under optimum allocation,

$$n_i = \frac{nN_iS_i}{\sum_{i=1}^{k} N_iS_i}$$

$$V_{opt}(\bar{y}_{st}) = \sum_{i=1}^{k}\left(\frac{1}{n_i} - \frac{1}{N_i}\right) w_i^2 S_i^2$$

$$= \sum_{i=1}^{k}\frac{w_i^2 S_i^2}{n_i} - \sum_{i=1}^{k}\frac{w_i^2 S_i^2}{N_i}$$

$$= \sum_{i=1}^{k}\left[w_i^2 S_i^2\left(\frac{\sum_{i=1}^{k} N_iS_i}{nN_iS_i}\right)\right] - \sum_{i=1}^{k}\frac{w_i^2 S_i^2}{N_i}$$

$$= \sum_{i=1}^{k}\left[\frac{1}{n}.\frac{N_iS_i}{N^2}\left(\sum_{i=1}^{k} N_iS_i\right)\right] - \sum_{i=1}^{k}\frac{w_i^2 S_i^2}{N_i}$$

$$= \frac{1}{n}\left(\sum_{i=1}^{k}\frac{N_iS_i}{N}\right)^2 - \sum_{i=1}^{k}\frac{w_i^2 S_i^2}{N_i}$$

$$= \frac{1}{n}\left(\sum_{i=1}^{k} w_iS_i\right)^2 - \frac{1}{N}\sum_{i=1}^{k} w_iS_i^2.$$

Comparison of variances of sample mean under SRS with stratified mean under proportional and optimal allocation:

(a) Proportional allocation

$$V_{SRS}(\bar{y}) = \frac{N-n}{Nn}S^2$$

$$V_{prop}(\bar{y}_{st}) = \frac{N-n}{Nn}\sum_{i=1}^{k}\frac{N_iS_i^2}{N}.$$

In order to compare $V_{SRS}(\bar{y})$ and $V_{prop}(\bar{y}_{st})$ first we attempt to express S^2 as a function of S_i^2.

Consider

$$(N-1)S^2 = \sum_{i=1}^{k}\sum_{j=1}^{N_i}(Y_{ij} - \bar{Y})^2$$

$$= \sum_{i=1}^{k} \sum_{j=1}^{N_i} \left[(Y_{ij} - \bar{Y}_i) + (\bar{Y}_i - \bar{Y}) \right]^2$$

$$= \sum_{i=1}^{k} \sum_{j=1}^{N_i} (Y_{ij} - \bar{Y}_i)^2 + \sum_{i=1}^{k} \sum_{j=1}^{N_i} (\bar{Y}_i - \bar{Y})^2$$

$$= \sum_{i=1}^{k} (N_i - 1) S_i^2 + \sum_{i=1}^{k} N_i (\bar{Y}_i - \bar{Y})^2$$

$$\frac{N-1}{N} S^2 = \sum_{i=1}^{k} \frac{N_i - 1}{N} S_i^2 + \sum_{i=1}^{k} \frac{N_i}{N} (\bar{Y}_i - \bar{Y})^2.$$

For simplification, we assume that N_i is large enough to permit the approximation

$$\frac{N_i - 1}{N_i} \approx 1 \text{ and } \frac{N-1}{N} \approx 1.$$

Thus

$$S^2 = \sum_{i=1}^{k} \frac{N_i}{N} S_i^2 + \sum_{i=1}^{k} \frac{N_i}{N} (\bar{Y}_i - \bar{Y})^2$$

or $\frac{N-n}{Nn} S^2 = \frac{N-n}{Nn} \sum_{i=1}^{k} \frac{N_i}{N} S_i^2 + \frac{N-n}{Nn} \sum_{i=1}^{k} \frac{N_i}{N} (\bar{Y}_i - \bar{Y})^2$ (Premultiply by $\frac{N-n}{Nn}$ on both sides)

$$Var_{SRS}(\bar{Y}) = V_{prop}(\bar{y}_{st}) + \frac{N-n}{Nn} \sum_{i=1}^{k} w_i (\bar{Y}_i - \bar{Y})^2$$

Since $\sum_{i=1}^{k} w_i (\bar{Y}_i - \bar{Y})^2 \geq 0,$

$$\Rightarrow Var_{prop}(\bar{y}_{st}) \leq Var_{SRS}(\bar{y}).$$

Larger gain in the difference is achieved when $\bar{Y}_i$ differs from $\bar{Y}$ more.

(b) Optimum allocation

$$V_{opt}(\bar{y}_{st}) = \frac{1}{n} \left(\sum_{i=1}^{k} w_i S_i \right)^2 - \frac{1}{N} \sum_{i=1}^{k} w_i S_i^2.$$

Consider

$$V_{prop}(\bar{y}_{st}) - V_{opt}(\bar{y}_{st}) = \left[\left(\frac{N-n}{Nn} \right) \sum_{i=1}^{k} w_i S_i^2 \right] - \left[\frac{1}{n} \left(\sum_{i=1}^{k} w_i S_i \right)^2 - \frac{1}{N} \sum_{i=1}^{k} w_i S_i^2 \right]$$

$$= \frac{1}{n} \left[\sum_{i=1}^{k} w_i S_i^2 - \left(\sum_{i=1}^{k} w_i S_i \right)^2 \right]$$

$$= \frac{1}{n}\sum_{i=1}^{k} w_i S_i^2 - \frac{1}{n}\bar{S}^2$$

$$= \frac{1}{n}\sum_{i=1}^{k} w_i (S_i - \bar{S})^2$$

where

$$\bar{S} = \sum_{i=1}^{k} w_i S_i$$

$$\Rightarrow Var_{prop}(\bar{y}_{st}) - Var_{opt}(\bar{y}_{st}) \geq 0$$

$$\text{or } Var_{opt}(\bar{y}_{st}) \leq Var_{prop}(\bar{y}_{st}).$$

Larger gain in efficiency is achieved when S_i differ from $\bar{S}$ more.

Combining the results in (a) and (b), we have

$$Var_{opt}(\bar{y}_{st}) \leq Var_{prop}(\bar{y}_{st}) \leq Var_{SRS}(\bar{y})$$

Estimate of Variance and Confidence Intervals

Under SRSWOR, an unbiased estimate of S_i^2 for the ith stratum (i = 1,2,...,k) is

$$s_i^2 = \frac{1}{n_i - 1}\sum_{j=1}^{n_i}(y_{ij} - \bar{y}_i)^2.$$

In stratified sampling,

$$Var(\bar{y}_{st}) = \sum_{i=1}^{k} w_i^2 \frac{N_i - n_i}{N_i n_i} S_i^2.$$

So an unbiased estimate of $Var(\bar{y}_{st})$ is

$$\widehat{Var}(\bar{y}_{st}) = \sum_{i=1}^{k} w_i^2 \frac{N_i - n_i}{N_i n_i} s_i^2$$

$$= \sum_{i=1}^{k} \frac{w_i^2 s_i^2}{n_i} - \sum_{i=1}^{k} \frac{w_i^2 s_i^2}{N_i}$$

$$= \sum_{i=1}^{k} \frac{w_i^2 s_i^2}{n_i} - \frac{1}{N}\sum_{i=1}^{k} w_i s_i^2.$$

The second term represents the reduction due to finite population correction.

The confidence limits of $\bar{Y}$ can be obtained as

$$\bar{y}_{st} \pm t\sqrt{\widehat{\text{Var}}(\bar{y}_{st})}$$

Assuming $\bar{y}_{st}$ is normally distributed and $\sqrt{\widehat{\text{Var}}(\bar{y}_{st})}$ is well determined so that t can be read from normal distribution tables. If only few degrees of freedom are provided by each stratum, then t values are obtained from the table of student's t-distribution.

The distribution of $\sqrt{\widehat{\text{Var}}(\bar{y}_{st})}$ is generally complex. An approximate method of assigning an effective number of degrees of freedom (n_e) to $\sqrt{\widehat{\text{Var}}(\bar{y}_{st})}$ is

$$n_e = \frac{\left(\sum_{i=1}^{k} g_i s_i^2\right)^2}{\sum_{i=1}^{k} \frac{g_i^2 s_i^4}{n_i - 1}}$$

where $g_i = \frac{N_i(N_i - n_i)}{n_i}$

and $Min(n_i - 1) \le n_e \le \sum_{i=1}^{k}(n_i - 1)$

Assuming y_{ij} are normally distributed.

Modification of Optimal Allocation

Sometimes in optimal allocation, the size of subsample exceeds the stratum size. In such a case,

- replace n_i by N_i and
- recompute the rest of n_i's by the revised allocation.

For example, if $n_i > N_1$ then take the revised n_i's as

$$\tilde{n}_1 = N_1$$

And

$$\tilde{n}_i = \frac{(n - N_1)w_i S_i}{\sum_{i=2}^{k} w_i S_i}; \quad i = 2, 3, \ldots, k$$

Provided $\tilde{n}_i \leq N_i$ for all i = 2,3,...,k.

Suppose in revised allocation, we find that $\tilde{n}_2 > N_i$ then the revised allocation would be

$$\tilde{n}_1 = N_1$$
$$\tilde{n}_2 = N_2$$
$$\tilde{n}_i = \frac{(n - N_1 - N_2) w_i S_i}{\sum_{i=3}^{k} w_i S_i}; i = 3, 4, \ldots, k$$

Provided $\tilde{n}_i < N_i$ for all i = 3,4,...,k.

We continue this process until every $\tilde{n}_i < N_i$

In such cases, the formula for minimum variance of $\bar{y}_{st}$ need to be modified as

$$Min\ Var(\bar{y}_{st}) = \frac{(\sum^* w_i S_i)^2}{n^*} - \frac{\sum^* w_i S_i^2}{N}$$

Where $\sum^*$ denotes the summation over the strata in which $\tilde{n}_i \leq N_i$ and n* is the revised total sample size in the strata.

Stratified Sampling for Proportions

If the characteristic under study is qualitative in nature, then its values will fall into one of the two mutually exclusive complementary classes C and C'. Ideally, only two strata are needed in which all the units can be divided depending on whether they belong to C or its complement C'. This is difficult to achieve in practice. So the strata are constructed such that the proportion in C varies as much as possible among strata.

Let

$P_i = \frac{A_i}{N_i}$: Proportion of units in C in i^{th} stratum

$p_i = \frac{a_i}{n_i}$: Proportion of units in C in the sample drawn from i^{th} stratum.

An estimate of population proportion based on stratified sampling is

$$p_{st} = \sum_{i=1}^{k} \frac{N_i p_i}{N}$$

which is based on the indicator variable

$$Y_{ij} = \begin{cases} 1 & \text{when } j^{th} \text{ unit belonging to } i^{th} \text{ stratum is in } C \\ 0 & \text{otherwise} \end{cases}$$

and $\overline{y}_{st} = p_{st}$.

Here $S_i^2 = \frac{N_i}{N_i - 1} P_i Q_i$

where $Q_i = 1 - P_i$.

Also $Var(\overline{y}_{st}) = \sum_{i=1}^{k} \frac{N_i - n_i}{N_i n_i} w_i^2 S_i^2$.

so $Var(p_{st}) = \frac{1}{N^2} \sum_{i=1}^{k} \frac{N_i^2 (N_i - n_i)}{N_i - 1} \frac{P_i Q_i}{n_i}$.

If the finite population correction can be ignored, then

$$Var(p_{st}) = \sum_{i=1}^{k} w_i^2 \frac{P_i Q_i}{n_i}.$$

If proportional allocation is used for n_i, then variance of p_{st} is

$$Var_{prop}(p_{st}) = \frac{N-n}{N} \frac{1}{Nn} \sum_{i=1}^{k} \frac{N_i^2 P_i Q_i}{N_i - 1}$$

$$= \frac{N-n}{Nn} \sum_{i=1}^{k} w_i P_i Q_i$$

and its estimate is

$$\widehat{Var}_{prop}(p_{st}) = \frac{N-n}{Nn} \sum_{i=1}^{k} w_i \frac{p_i q_i}{n_i - 1}.$$

The best choice of n_i such that it minimizes the variance for fixed total sample size is

$$n_i \propto N_i \sqrt{\frac{N_i P_i Q_i}{N_i - 1}}$$
$$= N_i \sqrt{P_i Q_i}.$$

Thus

$$n_i = n \frac{N_i \sqrt{P_i Q_i}}{\sum_{i=1}^{k} N_i \sqrt{P_i Q_i}}.$$

Similarly, the best choice of n_i such that the variance is minimum for fixed cost $C = C_0 + \sum_{i=1}^{k} C_i n_i$ is

$$n_i = \frac{nN_i \sqrt{\frac{P_i Q_i}{C_i}}}{\sum_{i=1}^{k} N_i \sqrt{\frac{P_i Q_i}{C_i}}}.$$

Estimation of the Gain in Precision Due to Stratification

An obvious question crops up that what is the advantage of stratifying a population in the sense that instead of using SRS, the population is divided into various strata? This is answered by estimating the variance of estimators of population mean under SRS (without stratification) and stratified sampling by evaluating

$$\frac{\widehat{Var}_{SRS}(\overline{y}) - \widehat{Var}(\overline{y}_{st})}{\widehat{Var}(\overline{y}_{st})}.$$

Since

$$Var_{SRS}(\overline{y}) = \frac{N-n}{Nn} S^2.$$

How to estimate S^2 based on a stratified sample?

Consider

$$(N-1)S^2 = \sum_{i=1}^{k}\sum_{j=1}^{N_i}(Y_{ij}-\overline{Y})^2$$

$$= \sum_{i=1}^{k}\sum_{j=1}^{N_i}\left[(Y_{ij}-\overline{Y}_i) + (\overline{Y}_i-\overline{Y})\right]^2$$

$$= \sum_{i=1}^{k}\sum_{j=1}^{N_i}(Y_{ij}-\overline{Y})^2 + \sum_{i=1}^{k}N_i(\overline{Y}_i-\overline{Y})^2$$

$$= \sum_{i=1}^{k}(N_i-1)S_i^2 + \sum_{i=1}^{k}N_i(\overline{Y}_i-\overline{Y})^2$$

$$= \sum_{i=1}^{k}(N_i-1)S_i^2 + N\left[\sum_{i=1}^{k}w_i\overline{Y}_i^2 - \overline{Y}^2\right].$$

In order to estimate S^2 , we need the estimates of $S_i^2, \overline{Y}_i^2$ *and* $\overline{Y}^2$ We consider their estimation one by one.

(I) For estimate of S_i^2 , we have

$$E(s_i^2) = S_i^2.$$

So $\hat{S}_i^2 = s_i^2$.

(II) For estimate of $\overline{Y}_i^2$, we know

$$Var(\overline{y}_i) = E(\overline{y}_i^2) - [E(\overline{y}_i)]^2$$
$$= E(\overline{y}_i^2) - \overline{Y}_i^2$$
$$\Rightarrow \overline{Y}_i^2 = E(\overline{y}_i^2) - Var(\overline{y}_i).$$

(III) For estimate of $\overline{Y}^2$, we know

$$Var(\overline{y}_{st}) = E(\overline{y}_{st}^2) - [E(\overline{y}_{st})]^2$$
$$= E(\overline{y}_{st}^2) - \overline{Y}^2$$
$$\Rightarrow \overline{Y}^2 = E(\overline{y}_{st}^2) - Var(\overline{y}_{st}) .$$

So an estimate of $\overline{Y}^2$ is

$$\hat{\overline{Y}}^2 = \overline{y}_{st}^2 - \widehat{Var}(\overline{y}_{st})$$
$$= \overline{y}_{st}^2 - \sum_{i=1}^{k}\left(\frac{N_i-n_i}{N_i n_i}\right)w_i^2 s_i^2.$$

Substituting these estimates in the expression $(n-1)S^2$ as follows, the estimate of S^2 is obtained as

$$(N-1)S^2 = \sum_{i=1}^{k}(N_i-1)S_i^2 + N\left[\sum_{i=1}^{k} w_i \bar{Y}_i^2 - \bar{Y}^2\right]$$

as $$\hat{S}^2 = \frac{1}{N-1}\sum_{i=1}^{k}(N_i-1)\hat{S}_i^2 + \frac{N}{N-1}\left[\sum_{i=1}^{k} w_i \hat{\bar{Y}}_i^2 - \hat{\bar{Y}}^2\right]$$

$$= \frac{1}{N-1}\left[\sum_{i=1}^{k}(N_i-1)s_i^2\right] + \frac{N}{N-1}\left[\left(\sum_{i=1}^{k} w_i\left(\bar{y}_i^2 - \left(\frac{N_i-n_i}{N_i n_i}\right)s_i^2\right)\right) - \left(\bar{y}_{st}^2 - \sum_{i=1}^{k}\frac{N_i-n_i}{N_i n_i}w_i^2 s_i^2\right)\right]$$

$$= \frac{1}{N-1}\left[\sum_{i=1}^{k}(N_i-1)s_i^2\right] + \frac{N}{N-1}\left[\sum_{i=1}^{k} w_i(\bar{y}_i - \bar{y}_{st})^2 - \sum_{i=1}^{k} w_i(1-w_i)\frac{N_i-n_i}{N_i n_i}s_i^2\right].$$

Thus

$$\widehat{Var}_{SRS}(\bar{y}) = \frac{N-n}{Nn}\hat{S}^2$$

$$= \frac{N-n}{N(N-1)n}\left[\sum_{i=1}^{k}(N_i-1)s_i^2\right] + \frac{N(N-n)}{nN(N-1)}\left[\sum_{i=1}^{k} w_i(\bar{y}_i - \bar{y}_{st})^2 - \sum_{i=1}^{k} w_i(1-w_i)\frac{N_i-n_i}{N_i n_i}s_i^2\right]$$

and

$$\widehat{Var}(\bar{y}_{st}) = \sum_{i=1}^{k}\frac{N_i-n_i}{N_i n_i}w_i^2 s_i^2.$$

Substituting these expressions in $\dfrac{\widehat{\mathrm{Var}}_{SRS}(\bar{y}) - \widehat{\mathrm{Var}}_{SRS}(\bar{y}_{st})}{\widehat{\mathrm{Var}}(\bar{y}_{st})}$,

the gain in efficiency due to stratification can be obtained.

If any other particular allocation is used, then substituting appropriate n_i under that allocation such gain can be estimated.

Interpenetrating Subsampling

Suppose a sample consists of two or more subsamples which are drawn according to the same sampling scheme. The samples are such that each subsample yields an estimate of parameter. Such subsamples are called interpenetrating subsamples.

The subsamples need not necessarily be independent. The assumption of independent subsamples helps in obtaining an unbiased estimate of the variance of the composite estimator. This is even helpful if the sample design is complicated and the expression for variance of the composite estimator is complex.

Let there be g independent interpenetrating subsamples and $t_1, t_2, \ldots, t_g$ be g unbiased estimators of parameter θ where $t_j(j=1,2,....,g)$ is based on j[th] interpenetrating subsample.

Then an unbiased estimator of θ is given by

$$\hat{\theta}=\frac{1}{g}\sum_{j=1}^{g}t_j=\bar{t}, \text{ say.}$$

$$\text{Then} \quad E(\hat{\theta})=E(\bar{t})=\theta$$

$$\text{and} \quad \widehat{Var}(\hat{\theta})=\widehat{Var}(\bar{t})=\frac{1}{g(g-1)}\sum_{j=1}^{g}(t_j-\bar{t})^2.$$

Note that

$$\begin{aligned}E\left[\widehat{Var}(\bar{t})\right]&=\frac{1}{g(g-1)}E\left[\sum_{j=1}^{g}(t_j-\theta)^2-g(\bar{t}-\theta)^2\right]\\&=\frac{1}{g(g-1)}E\left[\sum_{j=1}^{g}Var(t_j)-g\,Var(\bar{t})\right]\\&=\frac{1}{g(g-1)}(g^2-g)Var(\bar{t})\\&=Var(\bar{t}).\end{aligned}$$

If distribution of each estimator t_j is symmetric about θ, then the confidence interval of θ can be obtained by

$$P\left[Min(t_1,t_2,...,t_g)<\theta<Max(t_1,t_2,...,t_g)\right]=1-\left(\frac{1}{2}\right)^{g-1}.$$

Implementation of Interpenetrating Subsamples in Stratified Sampling

Consider the set up of stratified sampling. Suppose that each stratum provides an independent interpenetrating subsample. So based on each stratum, there are L independent interpenetrating subsamples drawn according to same sampling scheme.

Let $\hat{}_{ij(tot)}$ be the unbiased estimator of total of jth stratum based on the ith subsample, i = 1,2,...,L; j = 1,2,...,k.

An unbiased estimator of the jth stratum total is given by

$$\hat{Y}_{j(tot)}=\frac{1}{L}\sum_{i=1}^{J}\hat{Y}_{ij(tot)}$$

and an unbiased estimator of the variance of $\hat{Y}_{ij(tot)}$ is given by

$$\widehat{Var}(\hat{Y}_{j(tot)}) = \frac{1}{L(L-1)}\sum_{i=1}^{L}(\hat{Y}_{ij(tot)} - \hat{Y}_{j(tot)})^2.$$

Thus an unbiased estimator of population total Y_{tot} is

$$\hat{Y}_{tot} = \sum_{j=1}^{k}\hat{Y}_{j(tot)} = \frac{1}{k}\sum_{i=1}^{L}\sum_{j=1}^{k}\hat{Y}_{ij(tot)}$$

and unbiased estimator of its variance is

$$\widehat{Var}(\hat{Y}_{tot}) = \sum_{j=1}^{k}\widehat{Var}(\hat{Y}_{j(tot)}) = \frac{1}{L(L-1)}\sum_{i=1}^{L}\sum_{j=1}^{k}(\hat{Y}_{ij(tot)} - \hat{Y}_{j(tot)})^2.$$

Post Stratifications

Sometimes the stratum to which a unit belongs may be known after the field survey only. For example, the age of persons, their educational qualifications etc. can not be known in advance. In such cases, we adopt the post stratification procedure to increase the precision of the estimates.

In post stratification,

- Draw a sample by simple random sampling from the population and carry out the survey.
- After the completion of survey, stratify the sampling units to increase the precision of the estimates.

Assume that the stratum size N_i is fairly accurately known. Let

m_i : number of sampling units from i^{th} stratum, $i = 1,2,...,k$.

$$\sum_{i=1}^{k} m_i = n.$$

Note that m_i is a random variable (and that is why we are not using the symbol n_i as earlier).

Assume n is large enough or the stratification is such that the probability that some m_i =0 is negligibly small. In case, m_i =0 for some strata, two or more strata can be combined to make the sample size non- zero before evaluating the final estimates.

A post stratified estimator of population mean $\bar{Y}$ is

$$\overline{y}_{post} = \frac{1}{N}\sum_{i=1}^{k} N_i \overline{y}_i.$$

Now

$$E(\overline{y}_{post}) = \frac{1}{N} E\left[\sum_{i=1}^{k} N_i E(\overline{y}_i / m_1, m_2, ..., m_k)\right]$$

$$= \frac{1}{N} E\left[\sum_{i=1}^{k} N_i \overline{y}_i\right]$$

$$= \overline{Y}$$

$$Var(\overline{y}_{post}) = E\left[Var(\overline{y}_{post} \mid m_1, m_2, ..., m_k)\right] + Var\left[E(\overline{y}_{post} \mid m_1, m_2, ..., m_k)\right]$$

$$= E\left[\sum_{i=1}^{k} w_i^2 \left(\frac{1}{m_i} - \frac{1}{N_i}\right) S_i^2\right] + Var(\overline{Y})$$

$$= \sum_{i=1}^{k} w_i^2 \left[E\left(\frac{1}{m_i}\right) - \left(\frac{1}{N_i}\right)\right] S_i^2 \quad (Var(\overline{Y}) = 0).$$

To find $E\left(\frac{1}{m_i}\right) - \frac{1}{N_i}$, proceed as follows:

Consider the estimate of ratio based on ratio method of estimation as

$$\hat{R} = \frac{\overline{y}}{\overline{x}} = \frac{\sum_{j=1}^{n} y_j}{\sum_{j=1}^{n} x_j}, \qquad R = \frac{\overline{Y}}{\overline{X}} = \frac{\sum_{j=1}^{N} Y_j}{\sum_{j=1}^{N} X_j}.$$

We know that

$$E(\hat{R}) - R = \frac{N-n}{Nn} \cdot \frac{RS_X^2 - S_{XY}}{\overline{X}^2}.$$

Let

$$x_j = \begin{cases} 1 \text{ if } j^{th} \text{ unit belongs to } i^{th} \text{ stratum} \\ 0 \text{ otherwise} \end{cases}$$

and

$y_j = 1$ for all j = 1,2,...,N.

Then $R, \hat{R}$ and S_x^2 reduces to

$$\hat{R} = \frac{\sum_{j=1}^{n} y_j}{\sum_{j=1}^{n} x_j} = \frac{n}{n_i}$$

$$R = \frac{\sum_{j=1}^{N} y_j}{\sum_{j=1}^{N} x_j} = \frac{N}{N_i}$$

$$S_x^2 = \frac{1}{N-1}\left[\sum_{j=1}^{N} x_j^2 - N\overline{x}^2\right] = \frac{1}{N-1}\left[N_i - N\frac{N_i^2}{N^2}\right] = \frac{1}{N-1}\left(N_i - \frac{N_i^2}{N}\right)$$

$$S_{xy} = \frac{1}{N-1}\left[\sum_{j=1}^{N} x_j y_j - N\overline{x}\,\overline{y}\right] = \frac{1}{N-1}\left[N_i - \frac{N_i N}{N}\right] = 0.$$

Using these values in $E(\hat{R}) - R,$ we have

$$E(\hat{R}) - R = E\left(\frac{n}{n_i}\right) - \frac{N}{N_i} = \frac{N(N-n)(N-N_i)}{nN_i^2(N-1)}.$$

Thus

$$E\left(\frac{1}{n_i}\right) - \frac{1}{N_i} = \frac{N}{nN_i} + \frac{N(N-n)(N-N_i)}{n^2 N_i^2 (N-1)} - \frac{1}{N_i}$$

$$= \frac{(N-n)N}{n(N-1)N_i}\left(1 + \frac{N}{N_i n} - \frac{1}{n}\right).$$

Replacing m_i in place of n_i, we obtain

$$E\left(\frac{1}{m_i}\right) - \frac{1}{N_i} = \frac{(N-n)N}{n(N-1)N_i}\left(1 + \frac{N}{N_i n} - \frac{1}{n}\right).$$

Now substitute this in the expression of $Var(\bar{y}_{post})$ as

$$Var(\bar{y}_{post}) = \sum_{i=1}^{k} w_i^2 \left[E\left(\frac{1}{m_i}\right) - \frac{1}{N_i} \right] S_i^2$$

$$= \sum_{i=1}^{k} w_i^2 S_i^2 \left[\frac{N-n}{(N-1)n} \cdot \frac{N}{N_i} \left(1 + \frac{N}{nN_i} - \frac{1}{n} \right) \right]$$

$$= \frac{N-n}{n(N-1)} \sum_{i=1}^{k} w_i^2 S_i^2 \left[\frac{1}{w_i} \left(1 + \frac{1}{nw_i} - \frac{1}{n} \right) \right]$$

$$= \frac{N-n}{n^2(N-1)} \sum_{i=1}^{k} w_i S_i^2 \left[n - 1 + \frac{1}{w_i} \right]$$

$$= \frac{N-n}{n^2(N-1)} \sum_{i=1}^{k} (nw_i + 1 - w_i) S_i^2$$

$$= \frac{N-n}{n(N-1)} \sum_{i=1}^{k} w_i S_i^2 + \frac{N-n}{n^2(N-1)} \sum_{i=1}^{k} (1-w_i) S_i^2.$$

Assuming $N-1 \approx N.$

$$V(\bar{y}_{post}) = \frac{N-n}{Nn} \sum_{i=1}^{n} w_i^2 S_i^2 + \frac{N-n}{n^2 N} \sum_{i=1}^{n} (1-w_i) S_i^2$$

$$= V_{prop}(\bar{y}_{st}) + \frac{N-n}{Nn^2} \sum_{i=1}^{n} (1-w_i) S_i^2.$$

The second term is the contribution in the variance of $\bar{y}_{post}$ due to m_i's not being proportionately distributed.

If $S_i^2 \approx S_w^2$ say for all i, then the last term is

$$\frac{N-n}{Nn^2} \sum_{i=1}^{k} (1-w_i) S_i^2 = \frac{N-n}{Nn^2} S_w^2 (k-1) \quad \text{(Since } \sum_{i=1}^{k} w_i = 1)$$

$$= \left(\frac{k-1}{n}\right) \left(\frac{N-n}{Nn}\right) S_w^2$$

$$= \frac{k-1}{n} Var(\bar{y}_{st}).$$

The increase in variance over $Var_{prop}(\bar{y}_{st})$ is small if the average sample size ‾ – per stratum is reasonably large.

Thus a post stratification with a large sample produces an estimator which is almost as precise as an estimator in stratified sampling with proportional allocation.

Methods of Estimation

An important objective in any statistical estimation procedure is to obtain the estimators of parameters of interest with more precision. It is also well understood that incorporation of more information in the estimation procedure yields better estimators, provided the information is valid and proper. Use of such auxiliary information is made through the ratio method of estimation to obtain an improved estimator of population mean. In ratio method of estimation, auxiliary information on a variable is available which is linearly related to the variable under study and is utilized to estimate the population mean.

Let Y be the variable under study and X be any auxiliary variable which is correlated with Y. The observation x_i on X and y_i on Y are obtained for each sampling unit. The population mean $\bar{X}$ of X (or equivalently the population total X_{tot}) must be known. For example, x_i may be the values of y_i from

- some earlier completed census,
- some earlier surveys,
- some characteristic on which it is easy to obtain information etc.

For example, if y_i is the quantity of fruits produced in the ith plot, then x_i can be the area of ith plot or the production of fruit in the same plot in previous year.

Let $(x_1, y_1), (x_2, y_2), \ldots\ldots (x_n, y_n)$ be the random sample of size n on paired variable (X, Y) drawn, preferably by SRSWOR, from a population of size N. The ratio estimate of population mean is

$$\hat{\bar{Y}}_R = \frac{\bar{y}}{\bar{x}}\bar{X} = \hat{R}\bar{X}$$

assuming the population mean $\bar{X}$ is known. The ratio estimator of population total $Y_{tot} = \sum_{i=1}^{N} Y_i$

$$\hat{Y}_{R(tot)} = \frac{y_{tot}}{x_{tot}} X_{tot}$$

Where $X_{tot} = \sum_{i=1}^{N} X_i$ is the population total of X which is assumed to be known, $y_{tot} = \sum_{i=1}^{n} y_i$ and $x_{tot} = \sum_{i=1}^{n} x_i$ are the sample totals of Y and X respectively. The $\hat{Y}_{R(tot)}$ can be equivalently expressed as

$$\hat{Y}_{R(tot)} = \frac{\bar{y}}{\bar{x}} X_{tot}$$

$$= \hat{R}X_{tot}$$

Looking at the structure of ratio estimators, note that the ratio method estimates the relative change $\frac{Y_{tot}}{X_{tot}}$ that occurred after (x_i, y_i) were observed. It is clear that if the variation among the values of $\frac{y_i}{x_i}$ is nearly same for all i = 1,2,...,n then values of $\frac{y_{tot}}{x_{tot}}$ (or equivalently $\frac{\bar{y}}{\bar{x}}$) vary little from sample to sample and ratio estimate will be of high precision.

Bias and Mean Squared Error of Ratio Estimator

Assume that the random sample $(x_i, y_i), i = 1, 2, \ldots, n$ is drawn by SRSWOR and population mean $\bar{X}$ is known. Then

$$E(\hat{\bar{Y}}_R) = \frac{1}{\binom{N}{n}} \sum_{i=1}^{\binom{N}{n}} \frac{\bar{y}_i}{\bar{x}_i} \bar{X}$$

$$\neq \bar{Y} \text{ (in general)}.$$

Moreover it is difficult to find the exact expression for $E\left(\frac{\bar{y}}{\bar{x}}\right)$ and $E\left(\frac{\bar{y}^2}{\bar{x}^2}\right)$. So we approximate them and proceed as follows:

Let

$$\varepsilon_0 = \frac{\bar{y} - \bar{Y}}{\bar{Y}} \Rightarrow \bar{y} = (1 + \varepsilon_0)\bar{Y}$$

$$\varepsilon_1 = \frac{\bar{x} - \bar{X}}{\bar{X}} \Rightarrow \bar{x} = (1 + \varepsilon_1)\bar{X}.$$

Since SRSWOR is being followed, so

$$E(\varepsilon_0) = 0$$

$$E(\varepsilon_1) = 0$$

$$E(\varepsilon_0^2) = \frac{1}{\bar{Y}^2} E(\bar{y} - \bar{Y})^2$$

$$= \frac{1}{\bar{Y}^2} \frac{N - n}{Nn} S_Y^2$$

$$= \frac{f}{n} \frac{S_Y^2}{\bar{Y}^2}$$

$$= \frac{f}{n} C_Y^2$$

Where $f = \frac{N-n}{N}, S_Y^2 = \frac{1}{N-1}\sum_{i=1}^{N}(Y_i - \bar{Y})^2 and\ C_Y = \frac{S_Y}{\bar{Y}}$ is the coefficient of variation related to Y.

Similarly,

$$E(\varepsilon_1^2) = \frac{f}{n} C_X^2$$

$$E(\varepsilon_0 \varepsilon_1) = \frac{1}{\bar{X}\bar{Y}} E[(\bar{x} - \bar{X})(\bar{y} - \bar{Y})]$$

$$= \frac{1}{\bar{X}\bar{Y}} \frac{N-n}{Nn} \frac{1}{N-1} \sum_{i=1}^{N} (X_i - \bar{X})(Y_i - \bar{Y})$$

$$= \frac{1}{\bar{X}\bar{Y}} \frac{f}{n} S_{XY}$$

$$= \frac{1}{\bar{X}\bar{Y}} \frac{f}{n} \rho S_X S_Y$$

$$= \frac{f}{n} \rho \frac{S_X}{\bar{X}} \frac{S_Y}{\bar{Y}}$$

$$= \frac{f}{n} \rho C_X C_Y$$

Where $C_X = \frac{S_X}{\bar{X}}$ is the coefficient of variation related to X and ρ is the correlation coefficient between X and Y. Writing $\hat{\bar{Y}}_R$ in terms of $\varepsilon's$, we get

$$\hat{\bar{Y}}_R = \frac{\bar{y}}{\bar{x}} \bar{X}$$

$$= \frac{(1+\varepsilon_0)\bar{Y}}{(1+\varepsilon_1)\bar{X}} \bar{X}$$

$$= (1+\varepsilon_0)(1+\varepsilon_1)^{-1} \bar{Y}$$

Assuming $|\varepsilon_1| < 1,$ the term $(1+\varepsilon_1)^{-1}$ may be expanded as an infinite series and it would be convergent. Such assumption means that $\left|\frac{\bar{x} - \bar{X}}{\bar{X}}\right| < 1$ i.e., possible estimate $\bar{x}$ of population mean $\bar{X}$ lies between o and $2\bar{X}$, This is likely to hold true if the variation in $\bar{x}$ is not large. In order to ensures that variation in $\bar{x}$ is small, assume that the sample size n it is fairly large.

With this assumption,

$$\hat{\bar{Y}}_R = \bar{Y}(1+\varepsilon_0)(1-\varepsilon_1+\varepsilon_1^2-...)$$

$$=\bar{Y}(1+\varepsilon_0-\varepsilon_1+\varepsilon_1^2-\varepsilon_1\varepsilon_0+...).$$

So the estimation error of $\hat{\bar{Y}}_R$ is

$$\hat{\bar{Y}}_R-\bar{Y}=\bar{Y}(\varepsilon_0-\varepsilon_1+\varepsilon_1^2-\varepsilon_1\varepsilon_0+...).$$

In case, when sample size is large, then ε_0 *and* ε_1 are likely to be small quantities and so the terms involving second and higher powers of ε_0 *and* ε_1 would be negligibly small.

In such a case

$$\hat{\bar{Y}}_R-\bar{Y}\simeq\bar{Y}(\varepsilon_0-\varepsilon_1)$$

And

$$E(\hat{\bar{Y}}_R-\bar{Y})=0.$$

So the ratio estimator is an unbiased estimator of population mean upto the first order of approximation.

If we assume that only terms of ε_0 *and* ε_1 involving powers more than two are negligibly small (which is more realistic than assuming that powers more than one are negligibly small), then the estimation error of $\hat{\bar{Y}}_R$ can be approximated as

$$\hat{\bar{Y}}_R-\bar{Y}\simeq\bar{Y}(\varepsilon_0-\varepsilon_1+\varepsilon_1^2-\varepsilon_1\varepsilon_0)$$

And

$$E(\hat{\bar{Y}}_R-\bar{Y})=\bar{Y}\left(0-0+\frac{f}{n}C_X^2-\frac{f}{n}\rho C_X C_y\right)$$

$$Bias(\hat{\bar{Y}})=E(\hat{\bar{Y}}_R-\bar{Y})=\frac{f}{n}\bar{Y}C_X(C_X-\rho C_y)$$

upto second order of approximation, the bias generally decreases as the sample size grows large.

The bias of $\hat{\bar{Y}}_R$ is zero, i.e.,

$$Bias(\hat{\bar{Y}}_R) = 0$$

if $E(\varepsilon_1^2 - \varepsilon_0\varepsilon_1) = 0$

or if $\dfrac{Var(\bar{x})}{\bar{X}^2} - \dfrac{Cov(\bar{x},\bar{y})}{\bar{X}\bar{Y}} = 0$

or if $\dfrac{1}{\bar{X}^2}\left[Var(\bar{x}) - \dfrac{\bar{X}}{\bar{Y}}Cov(\bar{x},\bar{y})\right] = 0$

or if $Var(\bar{x}) - \dfrac{Cov(\bar{x},\bar{y})}{R} = 0$ (assuming $\bar{X} \neq 0$)

or if $R = \dfrac{\bar{Y}}{\bar{X}} = \dfrac{Cov(\bar{x},\bar{y})}{Var(\bar{x})}$

which is satisfied when the regression line of Y on X passes through origin.

Now, to find the mean squared error, consider

$$\begin{aligned} MSE(\hat{\bar{Y}}_R) &= E(\hat{\bar{Y}}_R - \bar{Y})^2 \\ &= E\left[\bar{Y}^2(\varepsilon_0 - \varepsilon_1 + \varepsilon_1^2 - \varepsilon_1\varepsilon_0 + \ldots)^2\right] \\ &\simeq E\left[\bar{Y}^2(\varepsilon_0^2 + \varepsilon_1^2 - 2\varepsilon_0\varepsilon_1)\right]. \end{aligned}$$

Under the assumption $|\varepsilon_1| < 1$ and the terms of ε_0 *and* ε_1 involving powers more than two are negligible small,

$$\begin{aligned} MSE(\hat{\bar{Y}}_R) &= \bar{Y}^2\left[\frac{f}{n}C_X^2 + \frac{f}{n}C_Y^2 - \frac{2f}{n}\rho C_X C_Y\right] \\ &= \frac{\bar{Y}^2 f}{n}\left[C_X^2 + C_Y^2 - 2\rho C_X C_y\right] \end{aligned}$$

up to the second order of approximation.

Eficiency of Ratio Estimator in Comparison to SRSWOR

Ratio estimator is better estimate of $\bar{Y}$ than sample mean based on SRSWOR if

$$MSE(\hat{\bar{Y}}_R) < Var_{SRS}(\bar{y})$$

$$\text{or if } \bar{Y}^2 \frac{f}{n}(C_X^2 + C_Y^2 - 2\rho C_X C_Y) < \bar{Y}^2 \frac{f}{n} C_Y^2$$

$$\text{or if } C_X^2 - 2\rho C_X C_Y < 0$$

$$\text{or if } \rho > \frac{1}{2}\frac{C_X}{C_Y}.$$

Thus ratio estimator is more efficient than sample mean based on SRSWOR if

$$\rho > \frac{1}{2}\frac{C_X}{C_Y} \quad if\ R > 0$$

$$and\ \rho > -\frac{1}{2}\frac{C_X}{C_Y} \quad if\ R < 0$$

It is clear from this expression that the success of ratio estimator depends on how close is the auxiliary information to the variable under study.

Upper Limit of Ratio Estimator

Consider

$$\begin{aligned} Cov(\hat{R}, \bar{x}) &= E(\hat{R}\bar{x}) - E(\hat{R})E(\bar{x}) \\ &= E\left(\frac{\bar{y}}{\bar{x}}\bar{x}\right) - E(\hat{R})E(\bar{x}) \\ &= \bar{Y} - E(\hat{R})\bar{X}. \end{aligned}$$

Thus

$$\begin{aligned} E(\hat{R}) &= \frac{\bar{Y}}{\bar{X}} - \frac{Cov(\hat{R}, \bar{x})}{\bar{X}} \\ &= R - \frac{Cov(\hat{R}, \bar{x})}{\bar{X}} \end{aligned}$$

$$\begin{aligned} Bias(\hat{R}) &= E(\hat{R}) - R \\ &= -\frac{Cov(\hat{R}, \bar{x})}{\bar{X}} \end{aligned}$$

$$= -\frac{\rho_{\hat{R},\bar{x}}\sigma_{\hat{R}}\sigma_{\bar{x}}}{\bar{X}}$$

Where $\rho_{\hat{R},\bar{x}}$ is the correlation between $\hat{R}$ *and* $\bar{x}$; $\sigma_{\hat{R}}$ *and* $\sigma_{\bar{x}}$ are the standard errors of $\hat{R}$ *and* $\bar{x}$ respectively.

Thus

$$\left|Bias(\hat{R})\right| = \frac{\left|-\rho_{\hat{R},\bar{x}}\right|\sigma_{\hat{R}}\sigma_{\bar{x}}}{\bar{X}}$$

$$\leq \frac{\sigma_{\hat{R}}\sigma_{\bar{x}}}{\bar{X}} \quad \left(\left|\rho_{\hat{R},\bar{x}}\right| \leq 1\right).$$

Assuming $\bar{X} > 0$.

Thus

$$\left|\frac{Bias(\hat{R})}{\sigma_{\hat{R}}}\right| \leq \frac{\sigma_{\bar{x}}}{\bar{X}}$$

$$or \left|\frac{Bias(\hat{R})}{\sigma_{\hat{R}}}\right| \leq C_X$$

Where C_X is the coefficient of variation of X. If $C_X < 0.1$, then the bias in $\hat{R}$ may be safely regarded as negligible in relation to standard error of $\hat{R}$.

Alternative form of $MSE(\hat{Y}_R)$

Consider

$$\sum_{i=1}^{N}(Y_i - RX_i)^2 = \sum_{i=1}^{N}\left[(Y_i - \bar{Y}) + (\bar{Y} - RX_i)\right]^2$$

$$= \sum_{i=1}^{N}\left[(Y_i - \bar{Y}) + R(X_i - \bar{X})\right]^2 \text{ (Using } \bar{Y} = R\bar{X}\text{)}$$

$$= \sum_{i=1}^{N}(Y_i - \bar{Y})^2 + R^2\sum_{i=1}^{N}(X_i - \bar{X})^2 - 2R\sum_{i=1}^{N}(X_i - \bar{X})(Y_i - \bar{Y})$$

$$\frac{1}{N-1}\sum_{i=1}^{N}(Y_i - RX_i)^2 = S_Y^2 + R^2 S_X^2 - 2RS_{XY}.$$

The MSE of $\hat{\bar{Y}}_R$ has already been derived which is now expressed again as follows:

$$MSE(\hat{\bar{Y}}_R) = \frac{f}{n}\bar{Y}^2(C_Y^2 + C_X^2 - 2\rho C_X C_Y)$$

$$= \frac{f}{n}\bar{Y}^2\left(\frac{S_Y^2}{\bar{Y}^2} + \frac{S_X^2}{\bar{X}^2} - 2\frac{S_{XY}}{\bar{X}\bar{Y}}\right)$$

$$= \frac{f}{n}\frac{\bar{Y}^2}{\bar{Y}^2}\left(S_Y^2 + \frac{\bar{Y}^2}{\bar{X}^2}S_X^2 - 2\frac{\bar{Y}}{\bar{X}}S_{XY}\right)$$

$$= \frac{f}{n}\left(S_Y^2 + R^2 S_X^2 - 2RS_{XY}\right)$$

$$= \frac{f}{n(N-1)}\sum_{i=1}^{N}(Y_i - RX_i)^2$$

$$= \frac{N-n}{nN(N-1)}\sum_{i=1}^{N}(Y_i - RX_i)^2.$$

Estimate of $MSE(\hat{Y}_R)$

Let $U_i = Y_i - RX_i, i = 1, 2, ..., N$ then MSE of $\hat{\bar{Y}}_R$ can be expressed as

Let $U_i = Y_i - RX_i, i = 1, 2, .., N$ then MSE of $\hat{\bar{Y}}_R$ can be expressed as

$$MSE(\hat{\bar{Y}}_R) = \frac{f}{n}\frac{1}{N-1}\sum_{i=1}^{N}(U_i - \bar{U})^2$$

$$= \frac{f}{n}S_U^2$$

where $S_U^2 = \frac{1}{N-1}\sum_{i=1}^{N}(U_i - \bar{U})^2.$

Based on this, a natural estimator of MSE $(\hat{\bar{Y}}_R)$ is

$$\widehat{MSE}(\hat{\bar{Y}}_R) = \frac{f}{n}s_u^2$$

where $s_u^2 = \frac{1}{n-1}\sum_{i=1}^{n}(u_i - \bar{u})^2$

$$= \frac{1}{n-1}\sum_{i=1}^{n}\left[(y_i - \bar{y}) - \hat{R}(x_i - \bar{x})\right]^2$$

$$= s_y^2 + \hat{R}^2 s_x^2 - 2\hat{R}s_{xy},$$

$$\hat{R} = \frac{\bar{y}}{\bar{x}}.$$

Based on the expression

$$MSE(\hat{\bar{Y}}_R) = \frac{f}{n(N-1)}\sum_{i=1}^{N}(Y_i - RX_i)^2,$$

an estimate of $MSE(\hat{\bar{Y}}_R)$ is

$$\widehat{MSE}\left(\hat{\bar{Y}}_R\right) = \frac{f}{n(n-1)}\sum_{i=1}^{n}(y_i - \hat{R}x_i)^2$$

$$= \frac{f}{n}(s_y^2 + \hat{R}^2 s_x^2 - 2\hat{R}s_{xy}).$$

Confidence Interval of Ratio Estimator

If the sample is large so that the normal approximation is applicable, then the $100(1-\alpha)\%$ confidence intervals of $\bar{Y}$ and R are

$$\left(\hat{\bar{Y}}_R - Z_{\frac{\alpha}{2}}\sqrt{\widehat{Var}(\hat{\bar{Y}}_R)}\ , \hat{\bar{Y}}_R + Z_{\frac{\alpha}{2}}\sqrt{\widehat{Var}(\hat{\bar{Y}}_R)}\right)$$

And

$$\left(\hat{R} - Z_{\frac{\alpha}{2}}\sqrt{\widehat{Var}(\hat{R})}\ , \hat{R} + Z_{\frac{\alpha}{2}}\sqrt{\widehat{Var}(\hat{R})}\right)$$

respectively where $Z_{\frac{\alpha}{2}}$ is the normal derivate to be chosen for given value of confidence coefficient $(1-\alpha)$.

If $(\bar{x}, \bar{y})$ follows a bivariate normal distributions, then $(\bar{Y} - R\bar{x})$ is normally distributed. If SRS is followed for drawing the sample, then assuming R is known, the statistic

$$\frac{\bar{y} - R\bar{x}}{\sqrt{\frac{N-n}{Nn}(s_y^2 + R^2 s_x^2 - 2Rs_{xy})}}$$

is approximately N(0,1).

Conditions Under Which the Ratio Estimate is Optimum

The ratio estimate $\hat{\bar{Y}}_R$ is best linear unbiased estimator of $\bar{Y}$ when

i. the relationship between y_i and x_i is linear passing through origin., i.e.

$$y_i = \beta x_i + e_i,$$

where e_i's are independent with $E(e_i|x_i) = 0$ and β is the slope parameter.

ii. this line is proportional to x_i, i.e.

$$Var(y_i|x_i) = E(e_i^2) = Cx_i$$

where C is constant.

Proof. Consider the linear estimate of β as $\hat{\beta} = \sum_{i=1}^{n} \ell_i y_i$ where $y_i = \beta x_i + e_i$

Then $\hat{\beta}$ is unbiased if $\bar{Y} = \beta \bar{X}$ because $E(y) = \beta \bar{X} + E(e_i|x_i)$.

If n sample values of x_i are kept fixed and then in repeated sampling

$$E(\hat{\beta}) = \sum_{i=1}^{n} \ell_i x_i \beta$$

and

$$Var(\hat{\beta}) = \sum_{i=1}^{n} \ell_i^2 Var(y_i|x_i) = C\sum_{i=1}^{n} \ell_i^2 x_i$$

$$So \;\; E(\hat{\beta}) = \beta \;\; when \;\; \sum_{i=1}^{n} \ell_i x_i = 1$$

Consider the minimization of $Var(y_i|x_i)$ subject to condition for unbiased estimator $\sum_{i=1}^{n} \ell_i x_i = 1$ using Lagrangian function.

Thus the Lagrangian function with Lagrangian multiplier λ is

$$\varphi = Var(y_i / x_i) - 2\lambda(\sum_{i=1}^{n} \ell_i x_i - 1)$$

$$= C\sum_{i=1}^{n} \ell_1^2 x_i - 2\lambda(\sum_{i=1}^{n} \ell_i x_i - 1).$$

Now

$$\frac{\partial \varphi}{\partial \ell_i} = 0 \Rightarrow \ell_i x_i = \lambda x_i,\ i = 1,2,..,n$$

$$\frac{\partial \varphi}{\partial \lambda} = 0 \Rightarrow \sum_{i=1}^{n} \ell_i x_i = 1.$$

Using

$$\sum_{i=1}^{n} \ell_i x_i = 1$$

$$\text{or} \quad \sum_{i=1}^{n} \lambda x_i = 1$$

$$\text{or} \quad \lambda = \frac{1}{n\overline{x}}.$$

$$\ell_i = \frac{1}{n\overline{x}}$$

and so

$$\hat{\beta} = \frac{\sum_{i=1}^{n} y_i}{n\overline{x}} = \frac{\overline{y}}{\overline{x}}.$$

Thus $\hat{\beta}$ is not only superior to $\overline{y}$ but also best in the class of linear and unbiased estimators.

Alternative Approach

This result can alternatively be derived as follows:

The ratio estimator $\hat{R} = \frac{\overline{y}}{\overline{x}}$ is the best linear unbiased estimator of $R = \frac{\overline{Y}}{\overline{X}}$ if the following two conditions hold:

i. For fixed x, $E(y) = \beta x$ i.e., the line of regression of y on x is a straight line passing through the origin.

ii. For fixed x, $Var(x) \propto x, i.e., Var(x) = \lambda x$ where λ is constant of proportionality

Proof: Let $\underline{y} = (y_1, y_2, \ldots, y_n)'$ and $\underline{x} = (x_1, x_2, \ldots, x_n)'$ be two vectors of observations on y's and x's . Hence for any fixed $\underline{x}$,

$$E(\underline{y}) = \beta \underline{x}$$

$$Var(\underline{y}) = \Omega = \lambda \; diag(x_1, x_2, ..., x_n)$$

Where $diag(x_1, x_2, ..., x_n)$ is the diagonal matrix with $x_1, x_2, ..., x_n$ as the diagonal elements.

The best linear unbiased estimator of β is obtained by minimizing

$$S^2 = (\underline{y} - \beta \underline{x})' \Omega^{-1} (\underline{y} - \beta \underline{x})$$

$$= \sum_{i=1}^{n} \frac{(y_i - \beta x_i)^2}{\lambda x_i}.$$

Solving

$$\frac{\partial S^2}{\partial \beta} = 0$$

$$\Rightarrow \sum_{i=1}^{n} (y_i - \hat{\beta} x_i) = 0$$

or $\hat{\beta} = \dfrac{\overline{y}}{\overline{x}} = \hat{R}$

Thus $\hat{R}$ is the best linear unbiased estimator of R. Consequently, $\hat{R}\overline{X} = \hat{\overline{Y}}_R$ is the best linear unbiased estimator of $\overline{Y}$.

Ratio Estimator in Stratified Sampling

Suppose a population of size N is divided into k strata. The objective is to estimate the population mean $\overline{Y}$ using ratio method of estimation.

In such situation, a random sample of size n_i is being drawn from i^{th} strata of size N_i on variable under study Y and auxiliary variable X using SRSWOR.

Let

y_{ij} : j^{th} observation on Y from i^{th} strata

x_{ij} : j^{th} observation on X from i^{th} strata i =1, 2,...,k; j = 1,2,...,n_i .

An estimator of $\overline{Y}$ based on the philosophy of stratified sampling can be derived in following two possible ways:

1. Separate ratio estimator

- Employ first the ratio method of estimation separately in each strata and obtain ratio estimator $\hat{\overline{Y}}_{R_i}$ $i = 1, 2, ..., k$ assuming the stratum mean $\overline{}$ to be known.
- Then combine all the estimates using weighted arithmetic mean.

This gives the separate ratio estimator as

$$\hat{\bar{Y}}_{Rs} = \sum_{i=1}^{k} \frac{N_i \hat{\bar{Y}}_{R_i}}{N}$$

$$= \sum_{i=1}^{k} w_i \hat{\bar{Y}}_{R_i}$$

$$= \sum_{i=1}^{k} w_i \frac{\bar{y}_i}{\bar{x}_i} \bar{X}_i$$

Where

$\bar{y}_i = \frac{1}{n_i} \sum_{j=1}^{n_i} y_{ij}$: sample mean of Y from ith strata

$\bar{x}_i = \frac{1}{n_i} \sum_{j=1}^{n_i} x_{ij}$: sample mean of X from ith strata

$\bar{X}_i = \frac{1}{N_i} \sum_{j=1}^{N_i} x_{ij}$: mean of all the units in ith strata

No assumption is made that the true ratio remains constant from stratum to stratum. It depends on information on each $\bar{X}_i$.

2. Combined ratio estimator:

- Find first the stratum mean of Y's and X's as

$$\bar{y}_{st} = \sum_{i=1}^{k} w_i \bar{y}_i$$

$$\bar{x}_{st} = \sum_{i=1}^{k} w_i \bar{x}_i.$$

- Then define the combined ratio estimator as

$$\hat{\bar{Y}}_{Rc} = \frac{\bar{y}_{st}}{\bar{x}_{st}} \bar{X}$$

Where $\bar{X}$ is the population mean of X based on all the $N = \sum_{i=1}^{N} N_i$ units. It does not depend on individual stratum units. It does not depend on information on each $\bar{X}_i$ but only on $\bar{X}$.

Jackknife Resampling

In statistics, the jackknife is a resampling technique especially useful for variance and bias estimation. The jackknife predates other common resampling methods such as the bootstrap. The jackknife estimator of a parameter is found by systematically leaving out each observation from a dataset and calculating the estimate and then finding the average of these calculations. Given a sample of size N, the jackknife estimate is found by aggregating the estimates of each $N-1$-sized sub-sample.

The jackknife technique was developed by Maurice Quenouille (1949, 1956). John Tukey (1958) expanded on the technique and proposed the name "jackknife" since, like a physical jack-knife (a compact folding knife), it is a rough-and-ready tool that can improvise a solution for a variety of problems even though specific problems may be more efficiently solved with a purpose-designed tool.

The jackknife is a linear approximation of the bootstrap.

Estimation

The jackknife estimate of a parameter can be found by estimating the parameter for each subsample omitting the *i*th observation to estimate the previously unknown value of a parameter (say $\bar{x}_i$).

$$\bar{x}_i = \frac{1}{n-1}\sum_{j \neq i}^{n} x_j$$

Variance Estimation

An estimate of the variance of an estimator can be calculated using the jackknife technique.

$$\text{Var}_{(\text{jackknife})} = \frac{n-1}{n}\sum_{i=1}^{n}(\bar{x}_i - \bar{x}_{(.)})^2$$

where $\bar{x}_i$ is the parameter estimate based on leaving out the *i*th observation, and $\bar{x}_{(.)} = \frac{1}{n}\sum_{i}^{n}\bar{x}_i$ is the estimator based on all of the subsamples.

Bias Estimation and Correction

The jackknife technique can be used to estimate the bias of an estimator calculated over the entire sample. Say $\hat{\theta}$ is the calculated estimator of the parameter of interest based on all n observations. Let

$$\hat{\theta}_{(.)} = \frac{1}{n}\sum_{i=1}^{n}\hat{\theta}_{(i)}$$

where $\hat{\theta}_{(i)}$ is the estimate of interest based on the sample with the ith observation removed, and $\hat{\theta}_{(.)}$ is the average of these "leave-one-out" estimates. The jackknife estimate of the bias of $\hat{\theta}$ is given by:

$$\widehat{\text{Bias}}_{(\theta)} = (n-1)(\hat{\theta}_{(.)} - \hat{\theta})$$

and the resulting bias-corrected jackknife estimate of θ is given by:

$$\hat{\theta}_{\text{Jack}} = n\hat{\theta} - (n-1)\hat{\theta}_{(.)}$$

This removes the bias in the special case that the bias is $O(N^{-1})$ and to $O(N^{-2})$ in other cases.

This provides an estimated correction of bias due to the estimation method. The jackknife does not correct for a biased sample.

Jackknife Method for Obtaining a Ratio Estimate with Lower Bias

Jackknife method, is used to get rid of the term of order 1/n from the bias of an estimator. Suppose the $E(\hat{R})$ can be expanded after ignoring finite population correction as

$$E(\hat{R}) = R + \frac{a_1}{n} + \frac{a_2}{n^2} + ...$$

Let n = mg and the sample is divided at random into g groups, each of size m. Then

$$E(g\hat{R}) = gR + \frac{ga_1}{gm} + \frac{ga_2}{g^2m^2} + ...$$

$$= gR + \frac{a_1}{m} + \frac{a_2}{gm^2} + ...$$

Let $\hat{R}_i^* = \frac{\sum^* y_i}{\sum^* x_i}$ where the $\sum^*$ denotes the summation over all values of the sample except the ith group. So $\hat{R}_i^*$ is based on a simple random sample of size m(g - 1), so we can express

$$E(\hat{R}_i^*) = R + \frac{a_1}{m(g-1)} + \frac{a_2}{m^2(g-1)^2} + ...$$

or

$$E\left[(g-1)\hat{R}_i^*\right] = (g-1)R + \frac{a_1}{m} + \frac{a_2}{m^2(g-1)} + ...$$

Thus

$$E\left[g\hat{R}-(g-1)\hat{R}_i^*\right]=R-\frac{a_2}{g(g-1)m^2}+...$$

or

$$E\left[g\hat{R}-(g-1)\hat{R}_i^*\right]=R-\frac{a_2}{n^2}\frac{g}{g-1}+...$$

Hence the bias of $[g\hat{R}-(g-1)\hat{R}_i^*]$ is of order $\frac{1}{n^2}$.

Now g estimates of this form can be obtained, one estimator for each group. Then the jackknife or Quenouille's estimator is the average of these of estimators

$$\hat{R}_Q=g\hat{R}-(g-1)\frac{\sum_{i=1}^{g}\hat{R}_i}{g}.$$

Product Method of Estimation

The ratio estimator is more efficient than the mean of a SRSWOR if $\rho>\frac{1}{2}.\frac{C_x}{C_y}$ provided R > o, which is usually the case. This shows that if auxiliary information is such that $\rho<-\frac{1}{2}\frac{C_x}{C_y}$, then we cannot use the ratio method of estimation to improve the sample mean as an estimator of population mean. So there is need of another type of estimator which also makes use of information on auxiliary variable x. Product estimator is an attempt in this direction.

The product estimator of the population mean $\overline{Y}$ is defined as

$$\hat{\overline{Y}}_p=\frac{\overline{y}\,\overline{x}}{\overline{X}}.$$

assuming the population mean $\overline{X}$ to be known

We now derive the bias and variance of $\hat{\overline{Y}}_p$.

Let

$$\varepsilon_0=\frac{\overline{y}-\overline{Y}}{\overline{Y}},\varepsilon_1=\frac{\overline{x}-\overline{X}}{\overline{X}},$$

(i) Bias of $\hat{\overline{Y}}_p$

We write $\hat{\overline{Y}}_p$ as

$$\hat{\bar{Y}}_p = \frac{\bar{y}\bar{x}}{\bar{X}} = \bar{Y}(1+\varepsilon_0)(1+\varepsilon_1)$$

$$= \bar{Y}(1+\varepsilon_0+\varepsilon_1+\varepsilon_0\varepsilon_1)$$

Taking expectation we obtain bias of $\hat{\bar{Y}}_p$ as

$$Bias(\hat{\bar{Y}}_p) = \frac{1}{\bar{X}}Cov(\bar{y},\bar{x}) = \frac{f}{n\bar{X}}S_{xy},$$

which shows that bias of $\hat{\bar{Y}}_p$ dcreases as n increases.

Bias of $\hat{\bar{Y}}_p$ can be estimated by

$$\widehat{\text{Bias}}\left(\hat{\bar{\text{Y}}}_\text{p}\right) = \frac{\text{f}}{\text{n}\bar{\text{X}}}\text{s}_{\text{xy}}$$

(ii) Variance of $\hat{\bar{Y}}_p$

Writing $\hat{\bar{Y}}_p$ in terms of ε_0 *and* ε_1, we find that the variance of the product estimator $\hat{\bar{Y}}_p$ up to second order of approximation is given by

$$Var(\hat{\bar{Y}}_p) = E(\hat{\bar{Y}}_p - \bar{Y})^2$$

$$= \bar{Y}^2 E(\varepsilon_0+\varepsilon_1+\varepsilon_0\varepsilon_1)^2$$

$$= \bar{Y}^2 E(\varepsilon_0^2+\varepsilon_1^2+2\varepsilon_0\varepsilon_1).$$

Here terms in $(\varepsilon_1, \varepsilon_0)$ of degrees greater than two are assumed to be negligible. Using the expected values we find that

$$Var(\hat{\bar{Y}}_p) = \frac{f}{n}[S_Y^2 + R^2 S_X^2 + 2RS_{XY}].$$

(iii) Estimation of variance of $\hat{\bar{Y}}_p$

The variance of $\hat{\bar{Y}}_p$ can be estimated by

$$\widehat{\text{Var}}(\hat{\bar{Y}}_p) = \frac{f}{n}[s_y^2 + r^2 s_x^2 + 2rs_{xy}]$$

Where $r = \bar{y}/\bar{x}$.

(iv) Comparison with SRSWOR:

From the variances of the mean of SRSWOR and the product estimator, we obtain

$$Var(\overline{y})_{SRS} - Var(\hat{\overline{Y}}_p) = -\frac{f}{n} RS_X (2\rho S_Y + RS_X)$$

which shows that $\hat{\overline{Y}}_p$ is more efficient than the simple mean $\overline{y}$ for

$$\rho < -\frac{1}{2}\frac{C_x}{C_y} \; if \; R > 0$$

and for

$$\rho > -\frac{1}{2}\frac{C_x}{C_y} \; if \; R < 0.$$

Multivariate Ratio Estimator

Let y be the study variable and $X_1, X_2, ..., X_p$ be p auxiliary variables assumed to be correlated with y. Further it is assumed that $X_1, X_2, ..., X_p$ are independent. Let $\overline{Y}, \overline{X}_1, \overline{X}_2, ...\overline{X}_p$ be the population means of the variables y, $X_1, X_2, ..., X_p$. We assume that a SRSWOR of size n is selected from the population of N units. The following notations will be used:

S_i^2 = the population mean sum of squares for the variate X_i,

s_i^2 = the sample mean sum of squares for the variate X_i,

S_0^2 = the population mean sum of squares for the study variable y,

s_0^2 = the sample mean sum of squares for the study variable y,

$C_i = \frac{S_i}{X_i}$ = coefficient of variation of the variate X_i

$C_0 = \frac{S_0}{Y}$ = coefficient of variation of the variate y,

$\rho_i = \frac{S_{iy}}{S_i S_0}$ = coefficient of correlation between y and X_i,

$\hat{\overline{Y}}_{Ri} = \frac{\overline{y}}{\overline{x}_i}\overline{X}_i$ = ratio estimator of $\overline{Y}$, based on X_i

where i=1,2,....,p . Then the multivariate ratio estimator of $\overline{Y}$ is given as follows

$$\hat{\overline{Y}}_{MR} = \sum_{i=1}^{p} w_i \hat{\overline{Y}}_{Ri}, \sum_{i=1}^{p} w_i = 1$$

$$= \overline{y} \sum_{i=1}^{p} w_i \frac{\overline{X}_i}{\overline{x}_i}.$$

(i) Bias of the multivariate ratio estimator :

The bias of $\hat{\overline{Y}}_{Ri}$ is

$$Bias(\hat{\overline{Y}}_{Ri}) = \frac{f}{n} \overline{Y}(C_i^2 - \rho_i C_i C_0).$$

The bias of $\hat{\overline{Y}}_{MR}$ is obtained as

$$Bias(\hat{\overline{Y}}_{MR}) = \sum_{i=1}^{p} w_i \frac{\overline{Y} f}{n}(C_i^2 - \rho_i C_i C_0)$$

$$= \frac{\overline{Y} f}{n} \sum_{i=1}^{p} w_i C_i (C_i - \rho_i C_0).$$

(ii) Variance of the multivariate ratio estimator :

The variance of $\hat{\overline{Y}}_{Ri}$ is given by

$$Var(\hat{\overline{Y}}_{Ri}) = \frac{f}{n} \overline{Y}^2 (C_0^2 + C_i^2 - 2\rho_i C_0 C_i).$$

The variance of $\hat{\overline{Y}}_{MR}$ is obtained as

$$Var(\hat{\overline{Y}}_{MR}) = \frac{f}{n} \overline{Y}^2 \sum_{i=1}^{p} w_i^2 (C_0^2 + C_i^2 - 2\rho_i C_0 C_i).$$

References

- Shahrokh Esfahani, Mohammad; Dougherty, Edward R. (2014). "Effect of separate sampling on classification accuracy". Bioinformatics. 30 (2): 242–250. PMID 24257187. doi:10.1093/bioinformatics/btt662

- Cameron, Adrian; Trivedi, Pravin K. (2005). Microeconometrics : methods and applications. Cambridge New York: Cambridge University Press. ISBN 9780521848053
- Efron, B.; Stein, C. (May 1981). "The Jackknife Estimate of Variance". The Annals of Statistics. 9 (3): 586–596. JSTOR 2240822. doi:10.1214/aos/1176345462
- Hunt, Neville; Tyrrell, Sidney (2001). "Stratified Sampling". Webpage at Coventry University. Archived from the original on 13 October 2013. Retrieved 12 July 2012
- Quenouille, M. H. (September 1949). "Problems in Plane Sampling". The Annals of Mathematical Statistics. 20 (3): 355–375. JSTOR 2236533. doi:10.1214/aoms/1177729989
- Efron, Bradley (1982). The jackknife, the bootstrap, and other resampling plans. Philadelphia, Pa: Society for Industrial and Applied Mathematics. ISBN 9781611970319
- Quenouille, M. H. (1956). "Notes on Bias in Estimation". Biometrika. 43 (3-4): 353–360. JSTOR 2332914. doi:10.1093/biomet/43.3-4.353
- Tukey, J. W. (1958). "Bias and confidence in not quite large samples". The Annals of Mathematical Statistics. 29: 614–623. doi:10.1214/aoms/1177706647

4

Cluster Sampling and Multistage Sampling

Cluster sampling is seen when the subsets of a set are mutually homogenous but the elements of each set have heterogeneous characteristics in a statistical population. In this, the whole set is divided into small subsets called clusters and simple random sample method is applied. The chapter closely examines the key concepts of cluster sampling to provide an extensive understanding of the subject.

Cluster Sampling

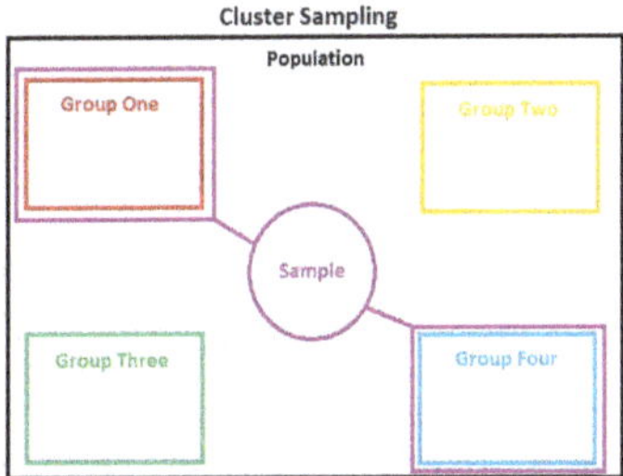

Cluster Sampling

Cluster sampling is a sampling plan used when mutually homogeneous yet internally heterogeneous groupings are evident in a statistical population. It is often used in marketing research. In this sampling plan, the total population is divided into these groups (known as clusters) and a simple random sample of the groups is selected. The elements in each cluster are then sampled. If all elements in each sampled cluster are sampled, then this is referred to as a "one-stage" cluster sampling plan. If a simple random subsample of elements is selected within each of these groups, this is referred to as a "two-stage" cluster sampling plan. A common motivation for cluster sampling is to reduce the total number of interviews and costs given the desired accuracy. For a fixed sample size, the expected random error is smaller when most of the variation in the population is present internally within the groups, and not between the groups.

Cluster Elements

The population within a cluster should ideally be as heterogeneous as possible, but there should be homogeneity between clusters. Each cluster should be a small-scale representation of the total population. The clusters should be mutually exclusive and collectively exhaustive. A random sampling technique is then used on any relevant clusters to choose which clusters to include in the study. In single-stage cluster sampling, all the elements from each of the selected clusters are sampled. In two-stage cluster sampling, a random sampling technique is applied to the elements from each of the selected clusters.

The main difference between cluster sampling and stratified sampling is that in cluster sampling the cluster is treated as the sampling unit so sampling is done on a population of clusters (at least in the first stage). In stratified sampling, the sampling is done on elements within each strata. In stratified sampling, a random sample is drawn from each of the strata, whereas in cluster sampling only the selected clusters are sampled. A common motivation of cluster sampling is to reduce costs by increasing sampling efficiency. This contrasts with stratified sampling where the motivation is to increase precision.

There is also multistage cluster sampling, where at least two stages are taken in selecting elements from clusters.

When Clusters Are of Different Sizes

Without modifying the estimated parameter, cluster sampling is unbiased when the clusters are approximately the same size. In this case, the parameter is computed by combining all the selected clusters. When the clusters are of different sizes, probability proportionate to size sampling is used. In this sampling plan, the probability of selecting a cluster is proportional to its size, so that a large clusters has a greater probability of selection than a small cluster. However, when clusters are selected with probability proportionate to size, the same number of interviews should be carried out in each sampled cluster so that each unit sampled has the same probability of selection.

Applications of Cluster Sampling

An example of cluster sampling is area sampling or geographical cluster sampling. Each clusters is a geographical area. Because a geographically dispersed population can be expensive to survey, greater economy than simple random sampling can be achieved by grouping several respondents within a local area into a cluster. It is usually necessary to increase the total sample size to achieve equivalent precision in the estimators, but cost savings may make such an increase in sample size feasible.

Cluster sampling is used to estimate high mortalities in cases such as wars, famines and natural disasters.

Advantage

- Can be cheaper than other sampling plans – e.g. fewer travel expenses, administration costs.

- Feasibility: This sampling plan takes large populations into account. Since these groups are so large, deploying any other sampling plan would be very costly.
- Economy: The regular two major concerns of expenditure, i.e., traveling and listing, are greatly reduced in this method. For example: Compiling research information about every household in a city would be very costly, whereas compiling information about various blocks of the city will be more economical. Here, traveling as well as listing efforts will be greatly reduced.
- Reduced variability: When estimates are being considered by any other method, reduced variability in results are observed. This may not be an ideal situation every time.

Major use: when sampling frame of all elements is not available we can resort only to the cluster sampling.

Disadvantage

- Higher sampling error, which can be expressed in the so-called "design effect", the ratio between the number of subjects in the cluster study and the number of subjects in an equally reliable, randomly sampled unclustered study.
- Biased samples: If the group in population that is chosen as a sample has a biased opinion, then the entire population is inferred to have the same opinion. This may not be the actual case.

Errors: The other probabilistic methods give fewer errors than this method. For this reason, it is discouraged for beginners.

More on Cluster Sampling

Two-stage Cluster Sampling

Two-stage cluster sampling, a simple case of multistage sampling, is obtained by selecting cluster samples in the first stage and then selecting sample of elements from every sampled cluster. Consider a population of N clusters in total. In the first stage, n clusters are selected using ordinary cluster sampling method. In the second stage, simple random sampling is usually used. It is used separately in every cluster and the numbers of elements selected from different clusters are not necessarily equal. The total number of clusters N, number of clusters selected n, and numbers of elements from selected clusters need to be pre-determined by the survey designer. Two-stage cluster sampling aims at minimizing survey costs and at the same time controlling the uncertainty related to estimates of interest. This method can be used in health and social sciences.

For instance, researchers used two-stage cluster sampling to generate a representative sample of the Iraqi population to conduct mortality surveys. Sampling in this method can be quicker and more reliable than other methods, which is why this method is now used frequently.

It is one of the basic assumptions in any sampling procedure that the population can be divided into a finite number of distinct and identifiable units, called sampling units. The smallest units into which the population can be divided are called elements of the population. The groups of such elements are called clusters.

In many practical situations and many types of populations, a list of elements is not available and so the use of an element as a sampling unit is not feasible. The method of cluster sampling or area sampling can be used in such situations.

In Cluster Sampling

- Divide the whole population into clusters according to some well defined rule.
- Treat the clusters as sampling units.
- Choose a sample of clusters according to some procedure.
- Carry out a complete enumeration of the selected clusters, i.e., collect information on all the sampling units available in selected clusters.

Area Sampling

In case, the entire area containing the populations is subdivided into smaller area segments and each element in the population is associated with one and only one such area segment, the procedure is called as area sampling.

Examples

- In a city, the list of all the individual persons staying in the houses may be difficult to obtain or even may be not available but a list of all the houses in the city may be available. So every individual person will be treated as sampling unit and every house will be a cluster.
- The list of all the agricultural farms in a village or a district may not be easily available but the list of village or districts are generally available. In this case, every farm is sampling unit and every village or district is the cluster.

Moreover, it is easier, faster, cheaper and convenient to collect information on clusters rather than on sampling units.

In both the examples, draw a sample of clusters from houses/villages and then collect the observations on all the sampling units available in the selected clusters.

Conditions Under which the Cluster Sampling is Used

Cluster sampling is preferred when

i. No reliable listing of elements is available and it is expensive to prepare it.

ii. Even if the list of elements is available, the location or identification of the units may be difficult.

iii. A necessary condition for the validity of this procedure is that every unit of the population under study must correspond to one and only one unit of the cluster so that the total number of sampling units in the frame may cover all the units of the population under study without any omission or duplication. When this condition is not satisfied, bias is introduced.

Open Segment and Closed Segment

It is not necessary that all the elements associated with an area segment need be located physically within its boundaries. For example, in the study of farms, the different fields of the same farm need not lie within the same area segment. Such a segment is called an open segment.

In a closed segment, the sum of the characteristic under study, i.e., area, livestock etc. for all the elements associated with the segment will account for all the area, livestock etc. within the segment.

Construction of Clusters

The clusters are constructed such that the sampling units are heterogeneous within the clusters and homogeneous among the clusters. The reason for this will become clear later. This is opposite to the construction of the strata in the stratified sampling.

There are two options to construct the clusters – equal size and unequal size. We discuss the estimation of population means and its variance in both the cases.

Case of Equal Clusters

- Suppose the population is divided into N clusters and each cluster is of size n.
- Select a sample of n clusters from N clusters by the method of SRS, generally WOR.

So

total population size = NM

total sample size = nM.

Let

y_{ij} : Value of the characteristic under study for the value of jth element (j = 1,2...,M) in the ith cluster (i = 1,2...,N)

$\bar{y}_i = \frac{1}{M}\sum_{j=1}^{M} y_{ij}$ mean per element of ith cluster.

Estimation of Population Mean

First select n clusters from N clusters by SRSWOR. Based on n clusters find the mean of each cluster separately based on all the units in every cluster. So we have the cluster means as $\bar{y}_1, \bar{y}_2, \ldots\ldots, \bar{y}_n$. Consider the mean of all such cluster means as an estimator of population mean as

$$\bar{y}_{cl} = \frac{1}{n}\sum_{i=1}^{n} \bar{y}_i$$

Bias

$$E(\bar{y}_{cl}) = \frac{1}{n}\sum_{i=1}^{n} E(\bar{y}_i)$$

$$= \frac{1}{n}\sum_{i=1}^{n} \bar{Y} \qquad \text{(Since SRS is used)}$$

$$= \bar{Y}$$

Thus $\bar{y}_{cl}$ is an unbiased estimator of $\bar{Y}$

Variance

The variance of $\bar{y}_{cl}$ can be derived on the same lines as deriving the variance of sample mean in SRSWOR. The only difference is that in SRSWOR, the sampling units are $y_1, y_2, \ldots\ldots, y_n$ whereas in case of $\bar{y}_{cl}$ the sampling units are $\bar{y}_1, \bar{y}_2, \ldots\ldots, \bar{y}_n$.

[Note that in case of SRSWOR, $Var(\bar{y}) = \frac{N-n}{Nn}S^2$ and $\widehat{\text{Var}}(\bar{y}) = \frac{N-n}{Nn}s^2$]

$$Var(\bar{y}_{cl}) = E\left(\bar{y}_{cl} - \bar{Y}\right)^2$$

$$= \frac{N-n}{Nn} S_b^2$$

Where $S_b^2 = \frac{1}{N-1}\sum_{i=1}^{N}\left(\bar{y}_{cl} - \bar{Y}\right)^2$ which is the mean sum of square between the cluster means in the population.

Estimate of Variance

Using again the philosophy of estimate of variance in case of SRSWOR, we can find

$$\widehat{\text{Var}}(\bar{y}_{cl}) = \frac{N-n}{Nn} s_b^2$$

Where $S_b^2 = \frac{1}{n-1}\sum_{i=1}^{n}\left(\bar{y}_i - \bar{y}_{cl}\right)^2$ is the mean sum of squares between cluster means in the sample.

Comparison with SRS

If an equivalent sample of *nM* units were to be selected from the population of *NM* units by SRSWOR, the variance of the mean per element would be

$$Var(\bar{y}_{nM}) = \frac{NM - nM}{NM} . \frac{S^2}{nM}$$

$$= \frac{f}{n} . \frac{S^2}{M}$$

$$\text{Where f} = \frac{N-n}{N} \text{and} \; S^2 = \frac{1}{NM-1}\sum_{i=1}^{N}\sum_{j=1}^{M}(y_{ij} - \bar{Y})^2.$$

$$\text{Also} \qquad Var(\bar{y}_{cl}) = \frac{N-n}{Nn} S_b^2$$

$$= \frac{f}{n} S_b^2.$$

Consider

$$(NM-1)S^2 = \sum_{i=1}^{N}\sum_{j=1}^{M}(y_{ij} - \bar{Y})^2$$

$$=\sum_{i=1}^{N}\sum_{j=1}^{M}[(y_{ij}-\bar{y}_i)+(\bar{y}_i-\bar{Y})]^2$$

$$=\sum_{i=1}^{N}\sum_{j=1}^{M}(y_{ij}-\bar{y}_i)^2+\sum_{i=1}^{N}\sum_{j=1}^{M}(\bar{y}_i-\bar{Y})^2$$

$$=N(M-1)\bar{S}_w^2+M(N-1)S_b^2$$

Where

$\bar{S}_w^2=\frac{1}{N}\sum_{i=1}^{N}S_i^2$ is the mean sum of squares within clusters in the population.

$S_i^2=\frac{1}{M-1}\sum_{j=1}^{M}(y_{ij}-\bar{y}_i)^2$ is the mean sum of squares for the ith cluster.

The efficiency of cluster sampling over SRSWOR is

$$E=\frac{Var(\bar{y}_{nM})}{Var(\bar{y}_{cl})}$$

$$=\frac{S^2}{MS_b^2}$$

$$=\frac{1}{(NM-1)}\left[\frac{N(M-1)}{M}\frac{\bar{S}_w^2}{S_b^2}+(N-1)\right].$$

Thus the relative efficiency increases when $\bar{S}_w^2$ is large and S_b^2 is small. So cluster sampling will be efficient if clusters are so formed that the variation between the cluster means is as small as possible while variation within the clusters is as large as possible.

Efficiency in Terms of Intra Class Correlation

The intra class correlation between the elements within a cluster is given by

$$\rho=\frac{E(y_{ij}-Y)(y_{ik}-\bar{Y})}{E(y_{ij}-\bar{Y})};\ -\frac{1}{M-1}\leq\rho\leq 1$$

$$=\frac{\frac{1}{MN(M-1)}\sum_{i=1}^{N}\sum_{j=1}^{M}\sum_{k(\neq j)=1}^{M}(y_{ij}-\bar{Y})(y_{ik}-\bar{Y})}{\frac{1}{MN}\sum_{i=1}^{N}\sum_{j=1}^{M}(y_{ij}-\bar{Y})^2}$$

$$\rho = \frac{E(y_{ij} - Y)(y_{ik} - \overline{Y})}{E(y_{ij} - \overline{Y})}; \ -\frac{1}{M-1} \le \rho \le 1$$

$$= \frac{\frac{1}{MN(M-1)} \sum_{i=1}^{N} \sum_{j=1}^{M} \sum_{k(\neq j)=1}^{M} (y_{ij} - \overline{Y})(y_{ik} - \overline{Y})}{\frac{1}{MN} \sum_{i=1}^{N} \sum_{j=1}^{M} (y_{ij} - \overline{Y})^2}$$

$$= \frac{\frac{1}{MN(M-1)} \sum_{i=1}^{N} \sum_{j=1}^{M} \sum_{k(\neq j)=1}^{M} (y_{ij} - \overline{Y})(y_{ik} - \overline{Y})}{\left(\frac{MN-1}{MN}\right) S^2}$$

$$= \frac{\sum_{i=1}^{N} \sum_{j=1}^{M} \sum_{k(\neq j)=1}^{M} (y_{ij} - \overline{Y})(y_{ik} - \overline{Y})}{(MN-1)(M-1)S^2}.$$

Consider

$$\sum_{i=1}^{N} (\overline{y}_i - \overline{Y})^2 = \sum_{i=1}^{N} \left[\frac{1}{M} \sum_{j=1}^{M} (y_{ij} - \overline{Y}) \right]^2$$

$$= \sum_{i=1}^{N} \left[\frac{1}{M^2} \sum_{j=1}^{M} (y_{ij} - \overline{Y})^2 + \frac{1}{M^2} \sum_{j=1}^{M} \sum_{k(\neq j)=1}^{M} (y_{ij} - \overline{Y})(y_{ik} - \overline{Y}) \right]^2$$

$$\Rightarrow \sum_{i=1}^{N} \sum_{j=1}^{M} \sum_{k(\neq j)=1}^{M} (y_{ij} - \overline{Y})(y_{ik} - \overline{Y}) = M^2 \sum_{i=1}^{N} (\overline{y}_i - \overline{Y})^2 - \sum_{i=1}^{N} \sum_{j=1}^{M} (y_{ij} - \overline{Y})^2$$

$$\text{or} \quad \rho(MN-1)(M-1)S^2 = M^2(N-1)S_b^2 - (NM-1)S^2$$

$$\text{or} \quad S_b^2 = \frac{(MN-1)}{M^2(N-1)} \left[1 + \rho(M-1)S^2\right]$$

The variance of $\overline{y}_{cl}$ now becomes

$$Var(\overline{y}_{cl}) = \frac{N-n}{N} S_b^2$$

$$= \frac{N-n}{Nn} \frac{MN-1}{N-1} \frac{S^2}{M^2} [1 + (M-1)\rho].$$

For large N, $\frac{MN-1}{MN} \simeq 1, \frac{N-n}{N} \simeq 1$ and so

$$Var(\overline{y}_{cl}) \simeq \frac{1}{n} \frac{S^2}{M} [1 + (1 + (M-1)\rho].$$

The variance of sample mean under SRSWOR for large N is

$$Var(\overline{y}_{nM}) \simeq \frac{S^2}{nM}.$$

The relative efficiency for large N is now given by

$$E = \frac{Var(\overline{y}_{nM})}{Var(\overline{y}_{cl})}$$

$$= \frac{\frac{S^2}{nM}}{\frac{S^2}{nM}[1+(M-1)\rho]}$$

$$= \frac{1}{1+(M-1)\rho}; 1 \le \rho \le -\frac{1}{(M-1)}$$

- If M = 1 then E = 1, i.e., SRS and cluster sampling are equally efficient. Each cluster will consist of one unit, i.e., SRS.

$$E > 1$$

$$or \ (M-1)\rho < 0$$

$$or \ \rho < 0.$$

- *f* If $\rho < 0$ then E = 1, i.e., there is no error which means that the units in each cluster are arranged randomly. So the sample is heterogeneous.
- In practice, ρ is usually positive and ρ decreases as M increases but the rate of decrease in ρ is much lower in comparison to the rate of increase in M. The situation that $\rho < 0$ is possible when the nearby units are grouped together to form cluster and which are completely enumerated.
- There are situations when $\rho < 0$.

Estimation of Relative Efficiency

The relative efficiency of cluster sampling relative to an equivalent SRSWOR is obtained as

$$E = \frac{S^2}{MS_b^2}$$

An estimator of E can be obtained by substituting the estimates of S^2 and S_b^2.

Since $\overline{y}_{cl} = \frac{1}{n}\sum_{i=1}^{n}\overline{y}_i$ is the mean of n means $\overline{y}_i$ from a population of N means $\overline{y}_i$, $i = 1, 2, \ldots\ldots, N$ which are drawn by SRSWOR, so from the theory of SRSWOR,

$$E(s_b^2) = E\left[\frac{1}{n-1}\sum_{i=1}^{n}(\overline{y}_i - \overline{y}_c)^2\right]$$

$$= \frac{1}{N-1}\sum_{i=1}^{N}(\bar{y}_i - \bar{Y})^2$$

$$= S_b^2$$

Thus s_b^2 is an unbiased estimator of S_b^2.

Since $s_w^2 = \frac{1}{n}\sum_{i=1}^{n} S_i^2$ is the mean of n mean sum of squares S_i^2 drawn from the population of N mean sums of squares $S_i^2, i = 1,2,\ldots\ldots,N,$ so it follows from the theory of SRSWOR that

$$E(s_w^2) = E\left[\frac{1}{n}\sum_{i=1}^{n} S_i^2\right]$$

$$= \frac{1}{N}\sum_{i=1}^{N} S_i^2$$

$$= \bar{S}_w^2.$$

Thus $\bar{s}_w^2$ is an unbiased estimator of $\bar{S}_w^2$.

Consider

$$S^2 = \frac{1}{MN-1}\sum_{i=1}^{N}\sum_{j=1}^{M}(y_{ij} - \bar{Y})^2$$

$$or\ (MN-1)S^2 = \sum_{i=1}^{N}\sum_{j=1}^{M}[(y_{ij} - \bar{y}_i) + (\bar{y}_i - \bar{Y})]^2$$

$$= \sum_{i=1}^{N}\sum_{j=1}^{M}[(y_{ij} - \bar{y}_i)^2 + (\bar{y}_i - \bar{Y})^2]$$

$$= \sum_{i=1}^{N}(M-1)S_i^2 + M(N-1)S_b^2$$

$$= N(M-1)\bar{S}_w^2 + M(N-1)S_b^2$$

An unbiased estimator of S^2 can be obtained as

$$\hat{S}^2 = \frac{1}{MN-1}[N(M-1)\bar{s}_w^2 + M(N-1)s_b^2]$$

So $\widehat{\text{Var}}(\bar{y}_{cl}) = \frac{N-n}{Nn} s_b^2$

$$\widehat{\mathrm{Var}}(\overline{y}_{nM}) = \frac{N-n}{Nn}\frac{\hat{S}^2}{M}$$

Where $s_b^2 = \frac{1}{n-1}\sum_{i=1}^{n}(\overline{y}_i - \overline{y}_{cl})^2$.

An estimate of efficiency $E = \frac{S^2}{MS_b^2}$ is

$$\hat{E} = \frac{N(M-1)\overline{s}_w^2 + M(N-1)s_b^2}{M(NM-1)s_b^2}.$$

If N is large so that $M(N-1) \simeq MN$ and $MN-1 \simeq MN$, then

$$E = \frac{1}{M} + \left(\frac{M-1}{M}\right)\frac{\overline{S}_w^2}{MS_b^2}$$

and its estimate is

$$\hat{E} = \frac{1}{M} + \left(\frac{M-1}{M}\right)\frac{\overline{s}_w^2}{Ms_b^2}$$

Estimation of a Proportion in Case of Equal Cluster

Now, we consider the problem of estimation of the proportion of units in the population having a specified attribute on the basis of a sample of clusters. Let this proportion be *P*.

Suppose that a sample of *n* clusters is drawn from *N* clusters with SRSWOR. Defining $y_{ij} = 1$ if the jth unit in the ith cluster belongs to the specified category (i.e. possessing the given attribute), we find that

$$\overline{y}_i = P_i,$$

$$\overline{Y} = \frac{1}{N}\sum_{i=1}^{N} P_i = P,$$

$$S_i^2 = \frac{MP_iQ_i}{(M-1)},$$

$$S_w^2 = \frac{M\sum_{i=1}^{N} P_iQ_i}{N(M-1)},$$

$$S^2 = \frac{NMPQ}{(NM-1)},$$

$$S_b^2 = \frac{1}{N-1}\sum_{i=1}^{N}(P_i - P)^2,$$

$$= \frac{1}{N-1}\left[\sum_{i=1}^{N} P_i^2 - NP^2\right]$$

$$= \frac{1}{(N-1)}\left[-\sum_{i=1}^{N} P_i(1-P_i) + \sum_{i=1}^{N} P_i \; - NP^2\right]$$

$$= \frac{1}{(N-1)}\left[NPQ - \sum_{i=1}^{N} P_i Q_i\right],$$

where P_i is the proportion of elements in the ith cluster, belonging to the specified category and $Q_i = 1 - P_i, i = 1,2,\ldots\ldots,N$ *and* $Q = 1 - P$ Then, using the result that $\overline{y}_{cl}$ is an unbiased estimator of $\overline{Y}$ we find that

$$\hat{P}_{cl} = \frac{1}{n}\sum_{i=1}^{n} P_i$$

is an unbiased estimator of P and

$$Var(\hat{P}_{cl}) = \left(\frac{N-n}{Nn}\right)\frac{\left[NPQ - \sum_{i=1}^{N} P_i Q_i\right]}{(N-1)}$$

This variance of $\hat{P}_{cl}$ can be expressed as

$$Var(\hat{P}_{cl}) = \frac{N-n}{N-1}\frac{PQ}{nM}[1 + (M-1)\rho],$$

where the value of ρ can be obtained as

$$\rho = \frac{M(N-1)S_b^2 - N\overline{S}_w^2}{(MN-1)}$$

by substituting $S_b^2, \overline{S}_w^2$ and S^2 in ρ, we obtain

$$\rho = 1 - \frac{M}{(M-1)}\frac{1}{N}\frac{\sum_{i=1}^{N} P_i Q_i}{PQ}.$$

The variance of $\hat{P}_c$ can be estimated unbiasedly by

$$\widehat{\text{Var}}\left(\hat{P}_c\right)=\frac{N-n}{nN}s_b^2$$

$$=\frac{N-n}{nN}\frac{1}{(n-1)}\sum_{i=1}^{n}(P_i-\hat{P}_c)^2$$

$$=\frac{N-n}{Nn(n-1)}\left[n\hat{P}_{cl}\hat{Q}_{cl}-\sum_{i=I}^{n}P_iQ_i\right]$$

Where $\hat{Q}_{cl}=I-\hat{P}_{cl}$. The efficiency of cluster sampling relative to SRSWOR is given by

$$E=\frac{M(N-1)}{(MN-1)}\frac{1}{\left[1+(M-1)\rho\right]}$$

$$=\frac{(N-1)}{NM-1}\frac{NPQ}{\left(NPQ-\sum_{i=1}^{N}P_iQ_i\right)}$$

If N is large, then $E\cong\frac{1}{M}$.

An estimator of the total number of elements belonging to a specified category is obtained by multiplying $\hat{P}_{cl}$ by NM, i.e. by $NM\hat{P}_{cl}$. The expressions of variance and variance estimator are obtained by multiplying the corresponding expressions for $\hat{P}_{cl}$ by N^2M^2.

Case of Unequal Clusters

In practice, the equal size of clusters are available only when planned. For example, in a screw manufacturing company, the packets of screws can be prepared such that every packet contains same number of screws. In real applications, it is hard to get clusters of equal size. For example, the villages with equal areas are difficult to find, the districts with same number of persons are difficult to find, the number of members in a household may not be same in each household in a given area.

Let there be N clusters and M_i be the size of ith cluster, let

$$M_0=\sum_{i=1}^{N}M_i$$

$$\bar{M}=\frac{1}{N}\sum_{i=1}^{N}M_i$$

$$\bar{y}_i=\frac{1}{M_i}\sum_{j=1}^{M_i}y_{ij}\ :\ \textit{mean of } i^{th} \textit{ cluster}$$

$$\bar{Y} = \frac{1}{M_0} \sum_{i=1}^{N} \sum_{j=1}^{M_i} y_{ij}$$

$$= \sum_{i=1}^{N} \frac{M_i}{M_0} \bar{y}_i$$

$$= \frac{1}{N} \sum_{i=1}^{N} \frac{M_i}{M_0} \bar{y}_i.$$

Suppose that n clusters are selected with SRSWOR and all elements in these selected clusters are surveyed. Assume that M_i's $(i = 1, 2, ..., N)$ are known.

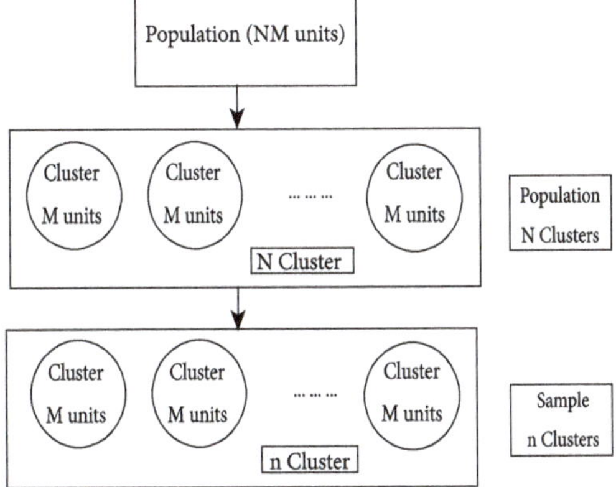

Based on this scheme, several estimators can be obtained to estimate the population mean. We consider four type of such estimators.

Mean of Cluster Means

Consider the simple arithmetic mean of the cluster means as

$$\bar{\bar{y}}_c = \frac{1}{n} \sum_{i=1}^{n} \bar{y}_i$$

$$E\left(\bar{\bar{y}}_c\right) = \frac{1}{N} \sum_{i=1}^{N} \bar{y}_i$$

$$\neq \bar{Y} \quad (where\ \bar{Y} = \sum_{i=1}^{N} \frac{M_i}{M_0} \bar{y}_i).$$

The bias of $\bar{\bar{y}}_c$ is

$$Bias\left(\bar{\bar{y}}_c\right) = E\left(\bar{\bar{y}}_c\right) - \bar{Y}$$

$$= \frac{1}{N}\sum_{i=1}^{N}\overline{y}_i - \sum_{i=1}^{N}\left(\frac{M_i}{M_0}\right)\overline{y}_i$$

$$= -\frac{1}{M_0}\left[\sum_{i=1}^{N}M_i\overline{y}_i - \frac{M_0}{N}\sum_{i=1}^{N}\overline{y}_i\right]$$

$$= -\frac{1}{M_0}\left[\sum_{i=1}^{N}M_i\overline{y}_i - \frac{\left(\sum_{i=1}^{N}M_i\right)\left(\sum_{i=1}^{N}\overline{y}_i\right)}{N}\right]$$

$$= -\frac{1}{M_0}\sum_{i=1}^{N}(M_i - \overline{M})(\overline{y}_i - \overline{Y})$$

$$= -\left(\frac{N-1}{M_0}\right)S_{m\overline{y}}$$

$Bias\left(\overline{\overline{y}}_c\right) = 0$ *if* M_i *and* $\overline{y}_i$ are uncorrelated.

The mean square error is

$$\text{MSE}\left(\overline{\overline{y}}_c\right) = Var\left(\overline{\overline{y}}_c\right) + \left[Bias\left(\overline{\overline{y}}_c\right)\right]^2$$

$$= \frac{N-n}{Nn}S_b^2 + \left(\frac{N-1}{M_0}\right)^2 S_{m\overline{y}}^2$$

Where

$$S_b^2 = \frac{1}{N-1}\sum_{i=1}^{N}(\overline{y}_i - \overline{Y})^2$$

$$S_{m\overline{y}} = \frac{1}{N-1}\sum_{i=1}^{N}(M_i - \overline{M})(\overline{y}_i - \overline{Y}).$$

An estimate of $Var(\overline{\overline{y}}_c)$ is

$$\widehat{\text{Var}}(\overline{\overline{y}}_c) = \frac{N-n}{Nn}s_b^2$$

Where $s_b^2 = \frac{1}{n-1}\sum_{i=1}^{n}(\overline{y}_c - \overline{\overline{y}}_c)^2.$

Weighted Mean of Cluster Means

Consider the arithmetic mean based on cluster total as

$$\bar{y}_c^* = \frac{1}{n\bar{M}}\sum_{i=1}^{n} M_i \bar{y}_i$$

$$E(\bar{y}_c^*) = \frac{1}{n}\sum_{i=1}^{n}\frac{1}{\bar{M}}E(\bar{y}_i M_i)$$

$$= \frac{n}{n}\frac{1}{M_0}\sum_{i=1}^{N} M_i \bar{y}_i$$

$$= \frac{1}{M_0}\sum_{i=1}^{N}\sum_{j=1}^{M_i} y_{ij}$$

$$= \bar{Y}.$$

Thus $\bar{y}_c^*$ is an unbiased estimator of $\bar{Y}$. The variance of $\bar{y}_c^*$ and its estimate are given by

$$Var(\bar{y}_c^*) = Var\left(\frac{1}{n}\sum_{i=1}^{n}\frac{M_i}{\bar{M}}\bar{y}_i\right)$$

$$= \frac{N-n}{Nn}S_b^{*2}$$

$$\widehat{\text{Var}}(\bar{y}_c^*) = \frac{N-n}{Nn}s_b^{*2}$$

Where

$$S_b^{*2} = \frac{1}{N-1}\sum_{i=1}^{N}\left(\frac{M_i}{\bar{M}}\bar{y}_i - \bar{Y}\right)^2$$

$$s_b^{*2} = \frac{1}{n-1}\sum_{i=1}^{n}\left(\frac{M_i}{\bar{M}}\bar{y}_i - \bar{y}_c^*\right)^2$$

$$E(s_b^{*2}) = S_b^{*2}.$$

Note that the expressions of variance of $\bar{y}_c^*$ and its estimate can be derived using directly the theory of SRSWOR as follows:

$$z_i = \frac{M_i}{\bar{M}}\bar{y}_i, then\ \bar{y}_c^* = \frac{1}{n}\sum_{i=1}^{n} z_i = \bar{z}.$$

Since SRSWOR is followed, so

$$Var(\overline{y}_c^*) = Var(\overline{z}) = \frac{N-n}{Nn}\frac{1}{N-1}\sum_{i=1}^{n}(z_i - \overline{Y})^2$$

$$= \frac{N-n}{Nn}\frac{1}{N-1}\sum_{i=1}^{N}\left(\frac{M_i}{\overline{M}}\overline{y}_i - \overline{Y}\right)^2$$

$$= \frac{N-n}{Nn}S_b^{*2}$$

Since

$$E(s_b^{*2}) = E\left[\frac{1}{n-1}\sum_{i=1}^{n}(z_i - \overline{z})^2\right]$$

$$= E\left[\frac{1}{n-1}\sum_{i=1}^{n}\left(\frac{M_i}{\overline{M}}(\overline{y}_i - \overline{y}_c^*)\right)^2\right]$$

$$= \frac{1}{N-1}\sum_{i=1}^{N}\left(\frac{M_i}{\overline{M}}(\overline{y}_i - \overline{Y})\right)^2$$

$$= S_b^{*2}.$$

So an unbiased estimator of variance can be easily derived.

Estimator based on Ratio Method of Estimation

Consider the weighted mean of the cluster means as

$$\overline{y}_c^{**} = \frac{\sum_{i=1}^{n} M_i \overline{y}_i}{\sum_{i=1}^{n} M_i}$$

It is easy to see that this estimator is a biased estimator of population mean. Before deriving its bias and mean squared error, we note that this estimator can be derived using the philosophy of ratio method of estimation. To see this, consider consider the study variable variable U_i and auxiliary auxiliary variable variable V_i as

$$U_i = \frac{M_i \overline{y}_i}{\overline{M}}$$

$$V_i = \frac{M_i}{\overline{M}} \; i = 1, 2, ..., N$$

$$\bar{V} = \frac{1}{N}\sum_{i=1}^{N} V_i = \frac{1}{N}\frac{\sum_{i=1}^{N} M_i}{\bar{M}} = 1$$

$$\bar{u} = \frac{1}{n}\sum_{i=1}^{n} u_i$$

$$\bar{v} = \frac{1}{n}\sum_{i=1}^{n} v_i.$$

The ratio estimator based on U and V is

$$\hat{\bar{Y}}_R = \frac{\bar{u}}{\bar{v}}\bar{V}$$

$$= \frac{\sum_{i=1}^{n} u_i}{\sum_{i=1}^{n} v_i}$$

$$= \frac{\sum_{i=1}^{n} \frac{M_i \bar{y}_i}{\bar{M}}}{\sum_{i=1}^{n} \frac{M_i}{\bar{M}}}$$

$$= \frac{\sum_{i=1}^{n} M_i \bar{y}_i}{\sum_{i=1}^{n} M_i}$$

Since the ratio estimator is biased, so $\bar{y}_c^{**}$ is also a biased estimator. The approximate bias and mean squared errors $\bar{y}_c^{**}$ can be derived directly by using the bias and MSE of ratio estimator. So using the results from the ratio method of estimation, the bias up to second order of approximation is given as follows

$$Bias(\bar{y}_c^{**}) = \frac{N-n}{Nn}\left(\frac{S_v^2}{\bar{V}^2} - \frac{S_{uv}}{\bar{U}\bar{V}}\right)\bar{U}$$

$$= \frac{N-n}{Nn}\left(S_v^2 - \frac{S_{uv}}{\bar{U}}\right)\bar{U}$$

where $\bar{U} = \frac{1}{N}\sum_{i=1}^{N} U_i = \frac{1}{N\bar{M}}\sum_{i=1}^{N} M_i \bar{y}_i$

$$S_v^2 = \frac{1}{N-1}\sum_{i=1}^{N}(V_i - \bar{V})^2$$

$$= \frac{1}{N-1}\sum_{i=1}^{N}\left(\frac{M_i}{\bar{M}} - 1\right)^2$$

$$S_{uv} = \frac{1}{N-1}\sum_{i=1}^{N}(U_i - \bar{U})(V_i - \bar{V})$$

$$= \frac{1}{N-1}\sum_{i=1}^{N}\left(\frac{M_i\bar{y}_i}{\bar{M}} - \frac{1}{N\bar{M}}\sum_{i=1}^{N}M_i\bar{y}_i\right)\left(\frac{M_i}{\bar{M}} - 1\right)$$

$$R_{uv} = \frac{\bar{U}}{\bar{V}} = \bar{U} = \frac{1}{N\bar{M}}\sum_{i=1}^{N}M_i\bar{y}_i.$$

The MSE of $\bar{y}_c^{**}$ up to second order of approximation can be obtained as follows:

$$MSE(\bar{y}_c^{**}) = \frac{N-n}{Nn}\left(S_u^2 + R^2 S_v^2 - 2RS_{uv}\right)$$

Where

$$S_u^2 = \frac{1}{N-1}\sum_{i=1}^{N}\left(\frac{M_i\bar{y}_i}{\bar{M}} - \frac{1}{N\bar{M}}\sum_{i=1}^{N}M_i\bar{y}_i\right)^2.$$

Alternatively,

$$MSE(\bar{y}_c^{**}) = \frac{N-n}{Nn}\frac{1}{N-1}\sum_{i=1}^{N}(U_i - R_{uv}V_i)^2$$

$$= \frac{N-n}{Nn}\frac{1}{N-1}\sum_{i=1}^{N}\left[\frac{M_i\bar{y}_i}{\bar{M}} - \left(\frac{1}{N\bar{M}}\sum_{i=1}^{N}M_i\bar{y}_i\right)\frac{M_i}{\bar{M}}\right]^2$$

$$= \frac{N-n}{Nn}\frac{1}{N-1}\sum_{i=1}^{N}\left(\frac{M_i}{\bar{M}}\right)^2\left[\bar{y}_i - \frac{\sum_{i=1}^{N}M_i\bar{y}_i}{N\bar{M}}\right]^2.$$

An estimator of MSE can be obtained as

$$\widehat{MSE}(\bar{y}_c^{**}) = \frac{N-n}{Nn}\frac{1}{n-1}\sum_{i=1}^{n}\left(\frac{M_i}{\bar{M}}\right)^2(\bar{y}_i - \bar{y}_c^{**})^2.$$

The estimator $\bar{\bar{y}}_c^{**}$ is biased but consistent.

4. Estimator based on unbiased ratio type estimation

Since $\bar{\bar{y}}_c = \frac{1}{n}\sum_{i=1}^{n}\bar{y}_i$ (where $\bar{y}_i = \frac{1}{M_i}\sum_{i=1}^{M_i} y_{ij}$) is a biased estimator of population mean and

$$Bias(\bar{\bar{y}}_c) = -\left(\frac{N-1}{M_0}\right)S_{m\bar{y}}$$
$$= -\left(\frac{N-1}{N\bar{M}}\right)S_{m\bar{y}}.$$

Since SRSWOR is used, so

$$S_{m\bar{y}} = \frac{1}{N-1}\sum_{i=1}^{n}(M_i - \bar{m})(\bar{y}_i - \bar{\bar{y}}_c),\ \bar{m} = \frac{1}{n}\sum_{i=1}^{n} M_i$$

is an unbiased estimator of

$$S_{m\bar{y}} = \frac{1}{n-1}\sum_{i=1}^{n}(M_i - \bar{M})(\bar{y}_i - \bar{Y}),$$

i.e.,

$$E(s_{m\bar{y}}) = S_{m\bar{y}}.$$

So it follow that

$$E(\bar{\bar{y}}_c) - \bar{Y} = -\left(\frac{N-1}{N\bar{M}}\right)E(s_{m\bar{y}}).$$

or

$$E\left[\bar{\bar{y}}_c + \left(\frac{N-1}{N\bar{M}}\right)s_{m\bar{y}}\right] = \bar{Y}$$

so

$$\bar{\bar{y}}_c^{**} = \bar{\bar{y}}_c + \left(\frac{N-1}{N\bar{M}}\right)s_{m\bar{y}}$$

is an unbiased estimator of the population mean $\bar{Y}$.

This estimator is based on unbiases ratio type esimator. This can be obtained by replacing the study variable (earlier y_i) by $\frac{M_i}{\bar{M}}\bar{y}_i$ and auxiliary variable (earlier x_i) by $\frac{M_i}{\bar{M}}$. The exact variance of this estimate is complicated and does not reduce to a simple form. The approximate variance upto the first orde of approximation is

$$Var\left(\bar{\bar{y}}_c^{**}\right) = \frac{1}{n(N-1)}\sum_{i=1}^{N}\left[\left(\frac{M_i}{\bar{M}}\bar{y}_i - \bar{Y}\right) - \left(\frac{1}{N\bar{M}}\sum_{i=1}^{N}\bar{y}_i\right)(M_i - \bar{M})\right]^2.$$

A consistent estimate of this variance is

$$\widehat{Var}\left(\bar{\bar{y}}_c^{**}\right) = \frac{1}{n(n-1)}\sum_{i=1}^{n}\left[\left(\frac{M_i}{\bar{M}}\bar{y}_i - \bar{y}_c\right) - \left(\frac{1}{n\bar{M}}\sum_{i=1}^{n}\bar{y}_i\right)\left(M_i - \frac{\sum_{i=1}^{n}M_i}{n}\right)\right]^2.$$

The variance of $\bar{\bar{y}}_c^{**}$ will be smaller than that of $\bar{y}_c^{**}$ (based on the ratio method of estimation) provided the regression coefficient of $\frac{M_i\bar{y}_i}{\bar{M}}$ on $\frac{M_i}{\bar{M}}$ is nearer to $\frac{1}{N}\sum_{i=1}^{N}\bar{y}_i$ than to $\frac{1}{M_0}\sum_{i=1}^{N}M_i\bar{y}_i$.

Comparison between SRS and cluster sampling

In case of unequal clusters, $\sum_{i=1}^{n}M_i$ is a random variable such that

$$E\left(\sum_{i=1}^{n}M_i\right) = n\bar{M}.$$

Now if a sample of size $n\bar{M}$ is drawn from a population of size $N\bar{M}$, then the variance of corresponding sample mean based on SRSWOR is

$$\begin{aligned} Var(\bar{y}_{SRS}) &= \frac{N\bar{M} - n\bar{M}}{N\bar{M}}\frac{S^2}{n\bar{M}} \\ &= \frac{N-n}{Nn}\frac{S^2}{\bar{M}}. \end{aligned}$$

This variance can be compared with any of the four proposed estimators.

For example, in case of

$$\bar{y}_c^* = \frac{1}{n\bar{M}}\sum_{i=1}^{n}M_i\bar{y}_i$$

$$Var(\bar{y}_c^*) = \frac{N-n}{Nn}S_b^{*2} = \frac{N-n}{Nn}\frac{1}{N-1}\sum_{i=1}^{N}\left(\frac{M_i}{\bar{M}}\bar{y}_i - \bar{Y}\right)^2.$$

The relative efficiency of $\overline{y}_c^{**}$ relative to SRS based sample mean

$$E = \frac{Var(\overline{y}_{SRS})}{Var(\overline{y}_c^*)}$$
$$= \frac{S^2}{\overline{M}S_b^{*2}}.$$

For $Var(\overline{y}_c^*) < Var(\overline{y}_{SRS})$, the variance between the clusters (S_b^{*2}) should be less. So the clusters should be formed in such a way that the variation between them is as samall as possible.

Sampling with Replacement and Unequal Probabilities (PPSWR)

In many practical situations, the cluster total for the study variable is likely to be positively correlated with the number of units in the cluster. In this situation, it is advantageous to select the clusters with probability proportional to the number of units in the cluster instead of with equal probability or to stratify the clusters according to their sizes and thn to draw a SRSWOR of clusters from each of the stratum. We consider here the case where clusters are selected with probability proportional to the number of units in the cluster and with replacement.

Suppose that n clusters are selected with PPSWR , the size being the number of units in the cluster. Here P_i is the probability of selection assigned to the ith cluster which is given by

$$P_i = \frac{M_i}{M_0} = \frac{M_i}{N\overline{M}}, \quad i = 1, 2, ..., N.$$

Consider the following estimator of the population mean:

$$\hat{\overline{Y}}_c = \frac{1}{n}\sum_{i=1}^{n} \overline{y}_i.$$

Then this estimator can be expressed as

$$\hat{\overline{Y}}_c = \frac{1}{n}\sum_{i=1}^{N} \alpha_i \overline{y}_i$$

where α_i denotes the number of times the ith cluster occurs in the sample. The random variables $\alpha_1, \alpha_2 ... \alpha_N$ follow a mutinomial probabbility distribution with

$$E(\alpha_i) = nP_i, \quad Var(\alpha_i) = nP_i(1 - P_i)$$
$$Cov(\alpha_i, \alpha_j) = -nP_iP_j, \quad i \neq j.$$

Hence,

$$E(\hat{\bar{Y}}_c) = \frac{1}{n}\sum_{i=1}^{N} E(\alpha_i)\bar{y}_i$$

$$= \frac{1}{n}\sum_{i=1}^{N} nP_i\bar{y}_i$$

$$= \sum_{i=1}^{N} \frac{M_i}{N\bar{M}}\bar{y}_i$$

$$= \frac{\sum_{i=1}^{N}\sum_{j=1}^{M_i} y_{ij}}{N\bar{M}} = \bar{Y}.$$

Thus $\hat{\bar{Y}}_c$ is an unbised estimator of $\bar{Y}$.

We now derive the variance of $\hat{\bar{Y}}_c$.

From $\hat{\bar{Y}}_c = \frac{1}{n}\sum_{i=1}^{N} \alpha_i\bar{y}_i$,

$$Var(\hat{\bar{Y}}_c) = \frac{1}{n^2}\left[\sum_{i=1}^{N} Var(\alpha_i)\bar{y}_i^2 + \sum_{i\neq j}^{N} Cov(\alpha_i,\alpha_j)\bar{y}_i\bar{y}_j\right]$$

$$= \frac{1}{n^2}\left[\sum_{i=1}^{N} P_i(1-P_i)\bar{y}_i^2 - \sum_{i\neq j}^{N} P_iP_j\bar{y}_i\bar{y}_j\right]$$

$$= \frac{1}{n^2}\left[\sum_{i=1}^{N} P_i\bar{y}_i^2 - \left(\sum_{i\neq j}^{N} P_i\bar{y}_i\right)^2\right]$$

$$= \frac{1}{n^2}\sum_{i=1}^{N} P_i\left(\bar{y}_i - \bar{Y}\right)^2$$

$$= \frac{1}{nN\bar{M}}\sum_{i=1}^{N} M_i(\bar{y}_i - \bar{Y})^2.$$

An unbiased estimator of the variance of $\hat{\bar{Y}}_c$ is

$$\widehat{Var}(\hat{\bar{Y}}_c) = \frac{1}{n(n-1)}\sum_{i=1}^{n}(\bar{y}_i - \hat{\bar{Y}}_c)^2$$

which can be seen to satisfy the unbisedness property as follows:

Consider

$$E\left[\frac{1}{n(n-1)}\sum_{i=1}^{n}(\bar{y}_i - \hat{\bar{Y}}_c)^2\right] = E\left[\frac{1}{n(n-1)}\left(\sum_{i=1}^{n}(\bar{y}_i^2 - n\hat{\bar{Y}}_c)^2\right)\right]$$

$$= \frac{1}{n(n-1)}\left[E\left(\sum_{i=1}^{n}\alpha_i\bar{y}_i^2\right) - nVar(\hat{\bar{Y}}_c) - n\bar{Y}^2\right]$$

where $E(\alpha_i) = nP_i, Var(\alpha_i) = nP_i(1-P_i), Cov(\alpha_i, \alpha_j) = -nP_iP_j, i \neq j$

$$E\left[\frac{1}{n(n-1)}\sum_{i=1}^{n}(\bar{y}_i - \hat{\bar{Y}}_c)^2\right] = \frac{1}{n(n-1)}\left[\sum_{i=1}^{N} n_i P_i \bar{y}_i^2 - n\frac{1}{n}\sum_{i=1}^{N} P_i(\bar{y}_i - \bar{Y})^2 - n\bar{Y}^2\right]$$

$$= \frac{1}{(n-1)}\left[\sum_{i=1}^{N} P_i(\bar{y}_i^2 - \bar{Y}^2) - \frac{1}{n}\sum_{i=1}^{N} P_i(\bar{y}_i - \bar{Y})^2\right]$$

$$= \frac{1}{(n-1)}\left[\sum_{i=1}^{N} P_i(\bar{y}_i - \bar{Y})^2 - \frac{1}{n}\sum_{i=1}^{N} P_i(\bar{y}_i - \bar{Y})^2\right]$$

$$= \frac{1}{(n-1)}\sum_{i=1}^{N} P_i(\bar{y}_i - \bar{Y})^2$$

$$= Var(\hat{\bar{Y}}_c).$$

Multistage Sampling

Multistage sampling refers to sampling plans where the sampling is carried out in stages using smaller and smaller sampling units at each stage.

Multistage sampling can be a complex form of cluster sampling because it is a type of sampling which involves dividing the population into groups (or clusters). Then, one or more clusters are chosen at random and everyone within the chosen cluster is sampled.

Using all the sample elements in all the selected clusters may be prohibitively expensive or unnecessary. Under these circumstances, multistage cluster sampling becomes useful. Instead of using all the elements contained in the selected clusters, the researcher randomly selects elements from each cluster. Constructing the clusters is the first stage. Deciding what elements within the cluster to use is the second stage. The technique is used frequently when a complete list of all members of the population does not exist and is inappropriate.

In some cases, several levels of cluster selection may be applied before the final sample elements are reached. For example, household surveys conducted by the Australian Bureau of Statistics begin by dividing metropolitan regions into 'collection districts' and selecting some of these collection districts (first stage). The selected collection districts are then divided into blocks, and blocks are chosen from within each selected collection district (second stage). Next, dwellings are listed within each selected block, and some of these dwellings are selected (third stage). This method makes it unnecessary to create a list of every dwelling in the region and necessary only for selected blocks. In remote areas, an additional stage of clustering is used, in order to reduce travel requirements.

Although cluster sampling and stratified sampling bear some superficial similarities, they are substantially different. In stratified sampling, a random sample is drawn from all the strata, where in cluster sampling only the selected clusters are studied, either in single- or multi-stage.

Advantages

- Cost and speed that the survey can be done in
- Convenience of finding the survey sample
- Normally more accurate than cluster sampling for the same size sample

Disadvantages

- Not as accurate as Simple Random Sample if the sample is the same size
- More testing is difficult to do

Two Stage Sampling

In cluster sampling, all the elements in the selected clusters are surveyed. Moreover, the efficiency in cluster sampling depends on size of the cluster. As the size increases, the efficiency decreases. It suggests that higher precision can be attained by distributing a given number of elements over a large number of clusters and then by taking a small number of clusters and enumerating all elements within them. This is achieved in subsampling.

In subsampling

- Divide the population into clusters.
- Select a sample of clusters [first stage]
- From each of the selected cluster, select a sample of specified number of elements [second stage]

The clusters which form the units of sampling at the first stage are called the first stage units and the units or group of units within clusters which form the unit of clusters are called the second stage units or subunits.

The procedure is generalized to three or more stages and is then termed as multistage sampling.

For example, in a crop survey

- villages are the first stage units,
- fields within the villages are the second stage units and
- plots within the fields are the third stage units.

In another example, to obtain a sample of fishes from a commercial fishery,

- first take a sample of boats and
- then take a sample of fishes from each selected boat.

Two Stage Sampling with Equal First Stage Units

Assume that

- population consists of NM elements.
- NM elements are grouped into N first stage units of M second stage units each, (i.e., N clusters, each cluster is of size M).
- Sample of n first stage units is selected (i.e., choose n clusters)
- Sample of m second stage units is selected from each selected first stage unit (i.e., choose m units from each cluster).
- Units at each stage are selected with SRSWOR.

Cluster sampling is a special case of two stage sampling in the sense that from a population of N clusters of equal size m = M, a sample of n clusters chosen.

If further M = m = 1, we get SRSWOR.

If n = N, we have the case of stratified sampling.

y_{ij} : Value of the characteristic under study for the jth second stage unit of the ith : first stage unit;

$$i = 1, 2, .., N; \; j = 1, 2, ..., m.$$

$\bar{Y}_i = \frac{1}{M}\sum_{j=1}^{m} y_{ij}$: mean per 2nd stage unit of ith 1st stage units in the population.

$\bar{Y} = \frac{1}{MN}\sum_{i=1}^{N}\sum_{j=1}^{M} y_{ij} = \frac{1}{N}\sum_{i=1}^{N} \bar{y}_i = \bar{Y}_{MN}$: mean per second stage unit in the population

$\bar{y}_i = \frac{1}{n}\sum_{j-1}^{m} y_{ij}$: mean per second stage unit in the ith first stage unit in the sample.

$\bar{y} = \frac{1}{mn}\sum_{i=1}^{n}\sum_{j=1}^{m} y_{ij} = \frac{1}{n}\sum_{i=1}^{n}\bar{y}_i = \bar{y}_{mn}$: mean per second stage in the sample.

Advantage

The principle advantage of two stage sampling is that it is more flexible than the one stage sampling. It reduces to one stage sampling when $m = M$ but unless this is the best choice of m, we have the opportunity of taking some smaller value that appears more efficient. As usual, this choice reduces to a balance between statistical precision and cost. When units of the first stage agree very closely, then consideration of precision suggests a small value of m. On the other hand, it is sometimes as cheap to measure the whole of a unit as to a sample. For example, when the unit is a household and a single respondent can give as accurate data as all the members of the household.

A pictorial scheme of two stage sampling scheme is as follows:

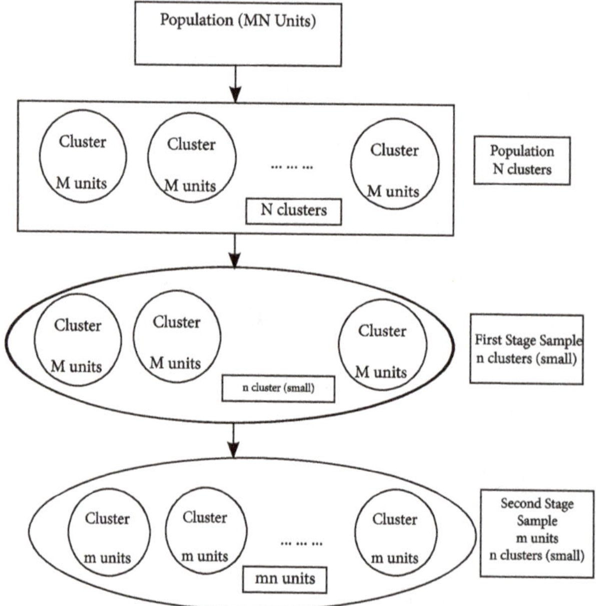

The expectations under two stage sampling scheme depend on the stages. For example, the expectation at second stage unit will be dependent on first stage unit in the sense that second stage unit will be in the sample provided it was selected in the first stage.

To Calculate the Average

- First average the estimator over all the second stage selections that can be drawn from a fixed set of n units that the plan selects.
- Then average over all the possible selections of n units by the plan.

In Case of Two Stage Sampling

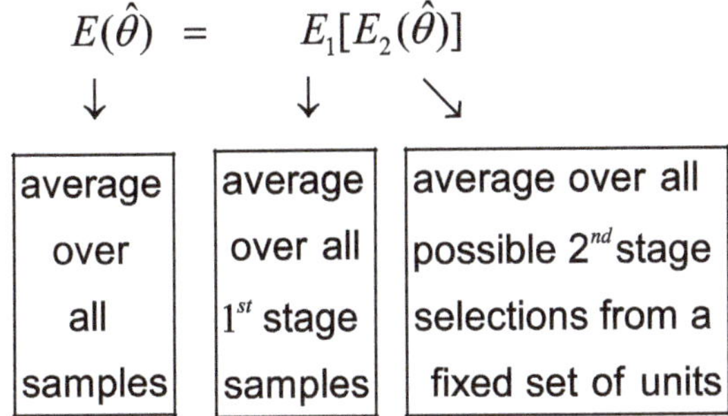

In Case of Three Stage Sampling

$$E(\hat{\theta}) = E_1[E_2\left\{E_3(\hat{\theta})\right\}].$$

To calculate the variance, we proceed as follows:

In case of two stage sampling,

$$Var(\hat{\theta}) = E(\hat{\theta} - \theta)^2$$

$$= E_1 E_2 (\hat{\theta} - \theta)^2.$$

Consider

$$E_2(\hat{\theta} - \theta)^2 = E_2(\hat{\theta}^2) - 2\theta E_2(\hat{\theta}) + \theta^2$$
$$= \left[\left\{E_2(\hat{\theta})\right\}^2 + V_2(\hat{\theta})\right] - 2\theta E_2(\hat{\theta}) + \theta^2.$$

Now average over first stage selection as

$$E_1 E_2(\hat{\theta} - \theta)^2 = E_1\left[E_2(\hat{\theta})\right]^2 + E_1\left[V_2(\hat{\theta})\right] - 2\theta E_1 E_2(\hat{\theta}) + E_1(\theta^2)$$
$$= E_1\left[E_1\left\{E_2(\hat{\theta})\right\}^2 - \theta^2\right] + E_1\left[V_2(\hat{\theta})\right]$$
$$Var(\hat{\theta}) = V_1\left[E_2(\hat{\theta})\right] + E_1\left[V_2(\hat{\theta})\right].$$

In case of three stage sampling,

$$Var(\hat{\theta}) = V_1\left[E_{\ 2}\left\{E_3(\hat{\theta})\right\}\right] + E_1\left[V_2\left\{E_3(\hat{\theta})\right\}\right] + E_1\left[E_2\left\{V_3(\hat{\theta})\right\}\right].$$

Estimation of Population Mean

Consider $\overline{y} = \overline{y}_{mn}$ as an estimator of the population mean $\overline{Y}$.

Bias

Consider

$$E(\overline{y}) = E_1\left[E_2(\overline{y}_{mn})\right]$$

$$= E_1\left[E_2(\overline{y}_{im} \mid i)\right] \quad \text{(as } 2^{nd} \text{ stage is dependent on } 1^{st} \text{ stage)}$$

$$= E_1\left[E_2(\overline{y}_{im} \mid i)\right] \quad \text{(as } y_i \text{ is unbiased for } \overline{Y}_i \text{ due to SRSWOR)}$$

$$= E_1\left[\frac{1}{n}\sum_{i=1}^{n}\overline{Y}_i\right]$$

$$= \frac{1}{N}\sum_{i=1}^{N}\overline{Y}_i$$

$$= \overline{Y}.$$

Thus $\overline{y}_{mn}$ is an unbiased of the population mean.

Variance

$$Var(\overline{y}) = E_1\left[V_2(\overline{y} \mid i)\right] + V_1\left[E_2(\overline{y} \mid i)\right]$$

$$= E_1\left[V_2\left\{\frac{1}{n}\sum_{i=1}^{n}\overline{y}_i \mid i\right\}\right] + V_1\left[E_2\left\{\frac{1}{n}\sum_{i=1}^{n}\overline{y}_i \mid i\right\}\right]$$

$$= E_1\left[\frac{1}{n^2}\sum_{i=1}^{n}V(\overline{y}_i \mid i)\right] + V_1\left[\frac{1}{n}\sum_{i=1}^{n}E_2(\overline{y}_i \mid i)\right]$$

$$= E_1\left[\frac{1}{n^2}\sum_{i=1}^{n}\left(\frac{1}{m}-\frac{1}{M}\right)S_i^2\right] + V_1\left[\frac{1}{n}\sum_{i=1}^{n}\overline{Y}_i\right]$$

$$= \frac{1}{n^2}\sum_{i=1}^{n}\left(\frac{1}{m}-\frac{1}{M}\right)E(S_i^2) \mid i + V_1(\overline{y}_c)$$

(where $\overline{y}_c$ is based on cluster means as in cluster sampling)

$$= \frac{1}{n^2}n\left(\frac{1}{m}-\frac{1}{M}\right)\overline{S}_w^2 + \frac{N-n}{Nn}S_b^2$$

$$= \frac{1}{n}\left(\frac{1}{m}-\frac{1}{M}\right)\overline{S}_w^2 + \left(\frac{1}{n}-\frac{1}{N}\right)S_b^2$$

where $\bar{S}_w^2 = \frac{1}{N}\sum_{i=1}^{N} S_i^2 = \frac{1}{N(M-1)}\sum_{i=1}^{N}\sum_{j=1}^{M}\left(Y_{ij} - \bar{Y}_i\right)^2$

$\bar{S}_b^2 = \frac{1}{N-1}\sum_{i=1}^{N}(\bar{Y}_i - \bar{Y})^2.$

Estimate of Variance

An unbiased estimator of variance of $\bar{y}$ can be obtained by replacing s_b^2 and $\bar{s}_w^2$ by their unbiased estimators in the expression of variance of $\bar{y}$

Consider an estimator of

$$\bar{S}_w^2 = \frac{1}{N}\sum_{i=1}^{N} S_i^2$$

where

$$S_i^2 = \frac{1}{M-1}\sum_{j=1}^{M}\left(y_{ij} - \bar{Y}_i\right)^2$$

and

$$\bar{s}_w^2 = \frac{1}{n}\sum_{i=1}^{n} s_i^2$$

$$s_i^2 = \frac{1}{m-1}\sum_{j=1}^{m}(y_{ij} - \bar{y}_i)^2.$$

so

$$E(\bar{s}_w^2) = E_1 E_2\left(\bar{s}_w^2 \mid i\right)$$

$$= E_1 E_2\left[\frac{1}{n}\sum_{i=1}^{n} s_i^2 \mid i\right]$$

$$= E_1 \frac{1}{n}\sum_{i=1}^{n}\left[E_2(s_i^2 \mid i)\right]$$

$$= E_1 \frac{1}{n}\sum_{i=1}^{n} S_i^2 \quad \text{(as SRSWOR is used)}$$

$$= \frac{1}{n}\sum_{i=1}^{n} E_1(S_i^2)$$

$$= \frac{1}{N}\sum_{i=1}^{N}\left[\frac{1}{N}\sum_{i=1}^{N} S_i^2\right]$$

$$= \frac{1}{N}\sum_{i=1}^{N} S_i^2$$

$$= \overline{S}_w^2$$

so $\overline{s}_w^2$ is an unbiased estimator of $\overline{S}_w^2$.

Consider

$$s_b^2 = \frac{1}{n-1}\sum_{i=1}^{n}(\overline{y}_i - \overline{y})^2$$

as an estimator of

$$S_b^2 = \frac{1}{N-1}\sum_{i=1}^{N}(\overline{Y}_i - \overline{Y})^2.$$

so

$$E(s_b^2) = \frac{1}{n-1}E\left[\sum_{i=1}^{n}(\overline{y}_i - \overline{y})^2\right]$$

$$(n-1)E(s_b^2) = E\left[\sum_{i=1}^{n}\overline{y}_i^2 - n\overline{y}^2\right]$$

$$= E\left[\sum_{i=1}^{n}\overline{y}_i^2\right] - nE(\overline{y}^2)$$

$$= E_1\left[E_2\left(\sum_{i=1}^{n}\overline{y}_i^2\right)\right] - n\left[Var(\overline{y}) + \{E(\overline{y})\}^2\right]$$

$$= E_1\left[\sum_{i=1}^{n}E_2(\overline{y}_i^2)\,|\,i)\right] - n\left[\left(\frac{1}{n}-\frac{1}{N}\right)S_b^2 + \left(\frac{1}{m}-\frac{1}{M}\right)\frac{1}{n}\overline{S}_w^2 + \overline{Y}^2\right]$$

$$= E_1\left[\sum_{i=1}^{n}\left\{Var(\overline{y}_i) + (E(\overline{y}_i))^2\right\}\right] - n\left[\left(\frac{1}{n}-\frac{1}{N}\right)S_b^2 + \left(\frac{1}{m}-\frac{1}{M}\right)\frac{1}{n}\overline{S}_w^2 + \overline{Y}^2\right]$$

$$= E_1\left[\sum_{i=1}^{n}\left\{\left(\frac{1}{m}-\frac{1}{M}\right)S_i^2 + \overline{Y}_i^2\right\}\right] - n\left[\left(\frac{1}{n}-\frac{1}{N}\right)S_b^2 + \left(\frac{1}{m}-\frac{1}{M}\right)\frac{1}{n}\overline{S}_w^2 + \overline{Y}^2\right]$$

$$= nE_1\left[\frac{1}{n}\left\{\sum_{i=1}^{n}\left(\frac{1}{m}-\frac{1}{M}\right)S_i^2 + \overline{Y}_i^2\right\}\right] - n\left[\left(\frac{1}{n}-\frac{1}{N}\right)S_b^2 + \left(\frac{1}{m}-\frac{1}{M}\right)\frac{1}{n}\overline{S}_w^2 + \overline{Y}^2\right]$$

$$= n\left[\left(\frac{1}{m}-\frac{1}{M}\right)\frac{1}{N}\sum_{i=1}^{N}S_i^2 + \frac{1}{N}\sum_{i=1}^{N}\overline{Y}_i^2\right] - n\left[\left(\frac{1}{n}-\frac{1}{N}\right)S_b^2 + \left(\frac{1}{m}-\frac{1}{M}\right)\frac{1}{n}\overline{S}_w^2 + \overline{Y}^2\right]$$

$$= n\left[\left(\frac{1}{m}-\frac{1}{M}\right)\overline{S}_w^2 + \frac{1}{N}\sum_{i=1}^{N}\overline{Y}_i^2\right] - n\left[\left(\frac{1}{n}-\frac{1}{N}\right)S_b^2 + \left(\frac{1}{m}-\frac{1}{M}\right)\frac{1}{n}\overline{S}_w^2 + \overline{Y}^2\right]$$

$$= (n-1)\left(\frac{1}{m}-\frac{1}{M}\right)\overline{S}_w^2 + \frac{n}{N}\sum_{i=1}^{N}\overline{Y}_i^2 - n\overline{Y}^2 - n\left(\frac{1}{n}-\frac{1}{N}\right)S_b^2$$

$$= (n-1)\left(\frac{1}{m}-\frac{1}{M}\right)\overline{S}_w^2 + \frac{n}{N}\left[\sum_{i=1}^{N}\overline{Y}_i^2 - N\overline{Y}^2\right] - n\left(\frac{1}{n}-\frac{1}{N}\right)S_b^2$$

$$= (n-1)\left(\frac{1}{m}-\frac{1}{M}\right)\overline{S}_w^2 + \frac{n}{N}(N-1)S_b^2 - n\left(\frac{1}{n}-\frac{1}{N}\right)S_b^2$$

$$= (n-1)\left(\frac{1}{m}-\frac{1}{M}\right)\overline{S}_w^2 + (n-1)S_b^2.$$

$$\Rightarrow E(s_b^2) = \left(\frac{1}{m}-\frac{1}{M}\right)\overline{S}_w^2 + S_b^2$$

or

$$E\left[s_b^2 - \left(\frac{1}{m}-\frac{1}{M}\right)\overline{s}_w^2\right] = S_b^2.$$

Thus

$$\widehat{Var}(\overline{y}) = \frac{1}{n}\left(\frac{1}{m}-\frac{1}{M}\right)\hat{\overline{S}}_\omega^2 + \left(\frac{1}{n}-\frac{1}{N}\right)\hat{S}_b^2$$

$$= \frac{1}{n}\left(\frac{1}{m}-\frac{1}{M}\right)\overline{s}_w^2 + \left(\frac{1}{n}-\frac{1}{N}\right)\left[s_b^2 - \left(\frac{1}{m}-\frac{1}{M}\right)\overline{s}_w^2\right]$$

$$= \frac{1}{N}\left(\frac{1}{m}-\frac{1}{M}\right)\overline{s}_w^2 + \left(\frac{1}{n}-\frac{1}{N}\right)s_b^2.$$

Allocation of Sample to the Two Stages: Equal First Stage Units

The variance of sample mean in the case of two stage sampling is

$$\widehat{\text{Var}}(\overline{y}) = \frac{1}{n}\left(\frac{1}{m}-\frac{1}{M}\right)\overline{S}_w^2 + \left(\frac{1}{n}-\frac{1}{N}\right)S_b^2.$$

It depends on S_b^2, $\overline{S}_w^2$, n and m. So the cost of survey of units in the two stage sample depends on n and m.

Case 1. When cost is fixed

We find the values of *n* and *m* so that the variance is minimum for given cost.

(i) When cost function is C = kmn

Let the cost of survey be proportional to sample size as

$C = knm$

where *C* is the total cost and *k* is constant

When cost is fixed as $C = C_0$. Substituting $m = \frac{C_o}{kn}$ in $Var(\overline{y})$, we get

$$Var(\overline{y}) = \frac{1}{n}\left[S_b^2 - \frac{\overline{S}_w^2}{M}\right] - \frac{S_b^2}{N} + \frac{1}{n}\frac{kn}{C_0}\overline{S}_w^2$$

$$= \frac{1}{n}\left(S_b^2 - \frac{\overline{S}_w^2}{M}\right) - \left(\frac{\overline{S}_w^2}{N} - \frac{k\overline{S}_w^2}{C_0}\right).$$

This variance is monotonic decreasing function of n if $\left(S_b^2 - \frac{\overline{S}_w^2}{M}\right) > 0$. The variance is minimum when n assumes maximum value, i.e., $\hat{n} = \frac{C_0}{k}$ corresponding to m=1. If $\left(S_b^2 - \frac{\overline{S}_w^2}{M}\right) < 0$ (i.e., intraclass correlation is negative for large N), then the variance is a monotonic increasing function of n. It reaches minimum when n assumes the minimum value, i.e., $\hat{n} = \frac{C_0}{km}$ (i.e., no subsampling).

(II) When cost function is $C = k_1 n + k_2 mn$

Let cost C be fixed as $C_0 = k_1 n + k_2 mn$ where k_1 and k_2 are positive constants. The terms denote the costs of per unit observations in the first and second stages. Minimize the variance of sample mean under the two stage with respect to m subject to the restriction $C_0 = k_1 n + k_2 mn$.

We have

$$C_0\left[Var(\overline{y}) + \frac{S_b^2}{N}\right] - k_1\left(S_b^2 - \frac{\overline{S}}{M}\right) + k_2\overline{S}_w^2 + mk_2\left(S_b^2 - \frac{\overline{S}_w^2}{M}\right) + \frac{k_1\overline{S}_w^2}{m}$$

When $\left(S_b^2 - \frac{\overline{S}}{M}\right)$>o, then

$$C_0\left[Var(\overline{y}) + \frac{S_b^2}{N}\right] = \left[\sqrt{k_1\left(S_b^2 - \frac{\overline{S}_w^2}{M}\right)} + \sqrt{k_2\overline{S}_w^2}\right]^2 + \left[\sqrt{mk_2\left(S_b^2 - \frac{\overline{S}_w^2}{M}\right)} - \sqrt{\frac{k_1\overline{S}_w^2}{m}}\right]^2$$

which is minimum when the second term of right hand side is zero. So we obtain

$$\hat{m} = \sqrt{\frac{k_1}{k_2}\frac{\overline{S}_w^2}{\left(S_b^2 - \frac{\overline{S}_w^2}{M}\right)}}.$$

The optimum n follows from $C_0 = k_1 n + k_2 mn$ as $\hat{n} = \dfrac{C_0}{k_1 + k_2 \hat{m}}$.

When $\left(S_b^2 - \dfrac{\bar{S}_w^2}{M} \right) \le 0$ then

$$C_0 \left[Var(\bar{y}) + \frac{S_b^2}{N} \right] = k_1 \left(S_b^2 - \frac{\bar{S}_w^2}{M} \right) + k_2 \bar{S}_w^2 + mk_2 \left(S_b^2 - \frac{\bar{S}_w^2}{M} \right) + \frac{k_1 \bar{S}_w^2}{m}$$

is minimum if m is the greatest attainable integer.

Hence in this case, when

$$C_0 \ge k_1 n + k_2 mn;\ \hat{m} = M \ and\ \hat{n} = \frac{C_0}{k_1 + k_2 M}.$$

If $C_0 \ge k_1 n + k_2 M;\ then\ \hat{m} = \dfrac{C_0 - k_1}{k_2}\ and\ \hat{n} = 1.$

If N is large, then

$$\bar{S}_w^2 \approx S^2 (1 - \rho)$$

$$\bar{S}_b^2 - \frac{\bar{S}_w^2}{M} \approx \rho S^2$$

$$\hat{m} \approx \sqrt{\frac{k_1}{k_2} \left(\frac{1}{\rho} - 1 \right)}.$$

Case 2: When variance is fixed

Now we find the sample sizes when variance is fixed, say as V_0.

$$V_0 = \frac{1}{N} \left(\frac{1}{m} - \frac{1}{M} \right) \bar{S}_w^2 + \left(\frac{1}{n} - \frac{1}{N} \right) S_b^2$$

$$\Rightarrow n = \frac{S_b^2 + \left(\frac{1}{m} - \frac{1}{M} \right) \bar{S}_w^2}{V_0 + \frac{S_b^2}{N}}.$$

So

$$C = kmn = km \left(\frac{S_b^2 - \frac{\bar{S}_w^2}{M}}{V_0 + \frac{\bar{S}_b^2}{N}} \right) + \frac{k \bar{S}_w^2}{V_0 + \frac{S_b^2}{N}}$$

If $\left(S_b^2 - \frac{\bar{S}_w^2}{M}\right) > 0; C$ attains minimum when m assumes the smallest integral value, i.e., 1.

If $\left(S_b^2 - \frac{\bar{S}_w^2}{M}\right) < 0; C$ attains minimum when $\hat{m} = M$.

Comparison of Two Stage Sampling with One Stage Sampling

One stage sampling procedures are comparable with two stage sampling procedures when either

(i)sampling mn elements in one single stage or

(ii)sampling $\frac{mn}{M}$ first stage units as cluster without sub-sampling.

We consider both the cases.

Case 1: Sampling mn Elements in One Single Stage

The variance of sample mean based on

- mn elements selected by SRSWOR (one stage) is given by

$$V(\bar{y}_{SRS}) = \left(\frac{1}{mn} - \frac{1}{MN}\right) S^2.$$

- two stage sampling is given by

$$V(\bar{y}_{TS}) = \frac{1}{n}\left(\frac{1}{m} - \frac{1}{M}\right) \bar{S}_w^2 + \left(\frac{1}{n} - \frac{1}{N}\right) S_b^2$$

The intraclass correlation coefficient is

$$\rho = \frac{M(N-1)S_b^2 - N\bar{S}_w^2}{(MN-1)S^2}; -\frac{1}{M-1} \le \rho \le 1$$

and using the identity

$$\sum_{i=1}^{N}\sum_{j=1}^{M}(y_{ij} - \bar{Y})^2 = \sum_{i=1}^{N}\sum_{j=1}^{M}(y_{ij} - \bar{Y}_i)^2 + \sum_{i=1}^{N}\sum_{j=1}^{M}(\bar{Y}_i - \bar{Y})^2$$

where $\bar{Y} = \frac{1}{MN}\sum_{i=1}^{N}\sum_{j=1}^{M} y_{ij}, \bar{Y}_i = \frac{1}{M}\sum_{j=1}^{M} y_{ij}$.

we have $(MN-1)S^2\rho = N(M-1)\bar{S}_w^2 + M(N-1)S_b^2$

and

$$(MN-1)S^2 = -N\overline{S}_w^2 + M(N-1)S_b^2$$

$$\Rightarrow S_b^2 = \frac{(MN-1)S^2}{M^2(N-1)}[1+(M-1)\rho] \quad \text{(Eliminating } \overline{S}_w^2)$$

$$\overline{S}_w^2 = \left(\frac{MN-1}{MN}\right)S^2(1-\rho).$$

Substituting S_b^2 and $\overline{S}_w^2$ in $Var(\overline{y}_{TS})$

$$V(\overline{y}_{TS}) = \left(\frac{MN-1}{MN}\right)\frac{S^2}{mn}\left[1-\frac{m(n-1)}{M(N-1)}+\rho\left\{\frac{N-n}{N-1}\frac{m}{M}(M-1)-\frac{M-m}{M}\right\}\right].$$

When subsampling rate $\frac{m}{M}$ is small, $MN-1 \approx MN$ and $M-1 \approx M$, then

$$\text{V}(\overline{\text{y}}_{SRS}) = \frac{S^2}{mn}$$

$$\text{V}(\overline{\text{y}}_{SRS}) = \frac{S^2}{mn}\left[1+\rho\left(\frac{N-n}{N-1}m-1\right)\right].$$

The relative efficiency of the two stage in relation to one stage sampling of SRSWOR is

$$RE = \frac{Var(\overline{\text{y}}_{TS})}{Var(\overline{\text{y}}_{SRS})} = 1+\rho\left(\frac{N-n}{N-1}m-1\right)$$

If $N-1 \approx N$ and finite population correction is ignorable,

Then $\frac{N-n}{N-1} \approx \frac{N-n}{N} \approx 1$,

and then $RE = 1+\rho(m-1)$

Case 2: Comparison with cluster sampling

Suppose a random sample of $\frac{mn}{M}$ clusters, without further subsampling is selected

The variance of the sample mean of equivalent nm/M clusters is

$$Var(\overline{y}_{cl}) = \left(\frac{M}{mn}-\frac{1}{n}\right)S_b^2$$

The variance of sample mean under the two stage sampling is

$$Var(\bar{y}_{TS}) = \frac{1}{n}\left(\frac{1}{m} - \frac{1}{M}\right)\bar{S}_w^2 + \left(\frac{1}{n} - \frac{1}{N}\right)S_b^2.$$

So $Var(\bar{y}_{cl})$ exceeds $Var(\bar{y}_{TS})$ by

$$\frac{1}{n}\left(\frac{M}{m} - 1\right)\left(S_b^2 - \frac{1}{M}\bar{S}_w^2\right)$$

which is approximately

$$\frac{1}{n}\left(\frac{M}{m} - 1\right)\rho S^2 \text{ for large } N \text{ and } \left(S_b^2 - \frac{\bar{S}_w^2}{M}\right) > 0$$

$$\text{where } S_b^2 = \frac{MN-1}{M(N-1)}\frac{S^2}{M}[1+\rho(M-1)]$$

$$\bar{S}_w^2 = \frac{MN-1}{MN}S^2(1-\rho).$$

So smaller the m /M, larger the reduction in the variance of two stage sample over a cluster sample.

When $\left(S_b^2 - \frac{\bar{S}_w^2}{M}\right) < 0$ then the subsampling will lead to loss in precision.

Two Stage Sampling with Unequal First Stage Units

Consider two stage sampling when the first stage units are of unequal size and SRSWOR is employed at each stage.

Let y_{ij}: value of jth second stage unit of the ith first stage unit.

M_i: number of second stage units in i[th] first stage unit.

$M_0 = \sum_{i=1}^{N} M_i$: total number of second stage units in the population.

m_i : number of second stage units to be selected from i[th] first stage units, if it is in the sample.

$m_0 = \sum_{i=1}^{n} m_i$: total number of second stage units in the sample.

$$\bar{y}_{i(m_i)} = \frac{1}{m_i}\sum_{j=1}^{m_i} y_{ij}$$

$$\bar{Y}_i = \frac{1}{M_i}\sum_{j=1}^{M_i} y_{ij}$$

$$\bar{Y} = \frac{1}{N}\sum_{i=1}^{N}\bar{y}_i = \bar{\bar{Y}}_N$$

$$\bar{\bar{Y}} = \frac{\sum_{i=1}^{N}\sum_{j=1}^{M_i} y_{ij}}{\sum_{i=1}^{N} M_i} = \frac{\sum_{i=1}^{N} M_i\bar{Y}_i}{\bar{M}N} = \frac{1}{N}\sum_{i=1}^{N} u_i\bar{Y}_i$$

$$u_i = \frac{M_i}{\bar{M}}$$

$$\bar{M} = \frac{1}{N}\sum_{i=1}^{N} M_i$$

The pictorial scheme of two stage sampling with unequal first stage units case is as follows:

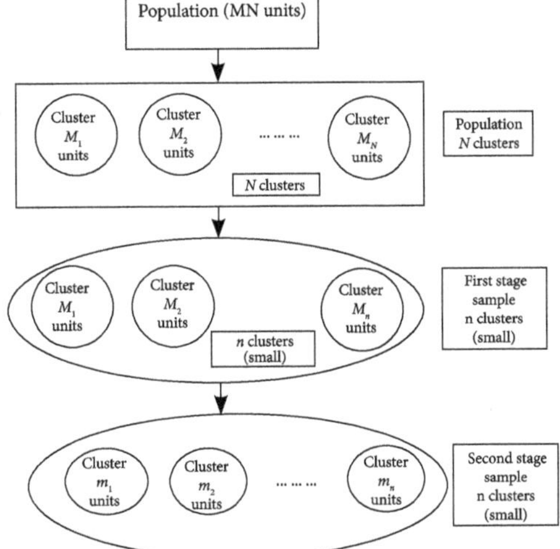

Now we consider different estimators for estimation of population mean.

1. Estimator based on first stage unit means in the sample:

$$\hat{\bar{Y}} = \bar{y}_{S2} = \frac{1}{n}\sum_{i=1}^{n}\bar{y}_{i(m_i)}$$

Bias

$$\begin{aligned} E(\bar{y}_{S2}) &= E\left[\frac{1}{n}\sum_{i=1}^{n}\bar{y}_{i(m_i)}\right] \\ &= E_1\left[\frac{1}{n}\sum_{i=1}^{n}E_2(\bar{y}_{i(m_i)})\right] \\ &= E_1\left[\frac{1}{n}\sum_{i=1}^{n}\bar{Y}_i\right] \quad \text{[Since a sample of size } m_i \text{ is selected out of } M_i \text{ units by SRSWOR]} \\ &= \frac{1}{N}\sum_{i=1}^{N}\bar{Y}_i \\ &= \bar{\bar{Y}}_N \\ &\neq \bar{\bar{Y}}. \end{aligned}$$

So $\bar{y}_{S2}$ is a biased estimator of $\bar{Y}$ and its bias is given by

$$\begin{aligned}Bias\ (\bar{y}_{S2}) &= E(\bar{y}_{S2})-\bar{Y}\\ &= \frac{1}{N}\sum_{i=1}^{N}\bar{Y}_i-\frac{1}{N\bar{M}}\sum_{i=1}^{N}M_i\bar{Y}_i\\ &= -\frac{1}{N\bar{M}}\left[\sum_{i=1}^{N}M_i\bar{Y}_i-\frac{1}{N}\left(\sum_{i=1}^{N}\bar{Y}_i\right)\left(\sum_{i=1}^{N}M_i\right)\right]\\ &= \frac{1}{N\bar{M}}\sum_{i=1}^{N}(M_i-\bar{M})(\bar{Y}_i-\bar{\bar{Y}}_N).\end{aligned}$$

This bias can be estimated by

$$\widehat{Bias}(\bar{y}_{S2}) = -\frac{N-1}{N\bar{M}(n-1)}\sum_{i=1}^{n}(M_i-\bar{m})(\bar{y}_{i(mi)}-\bar{y}_{S2})$$

which can be seen as follows

$$\begin{aligned}E\left[\widehat{Bias}(\bar{y}_{S2})\right] &= -\frac{N-1}{N\bar{M}}E_1\left[\frac{1}{n-1}\sum_{i=1}^{n}E_2\left\{(M_i-\bar{m})(\bar{y}_{i(mi)}-\bar{y}_{S2})/n\right\}\right]\\ &= -\frac{N-1}{N\bar{M}}E\left[\frac{1}{n-1}\sum_{i=1}^{n}(M_i-\bar{m})(\bar{Y}_i-\bar{\bar{y}}_n)\right]\\ &= -\frac{1}{N\bar{M}}\sum_{i=1}^{N}(M_i-\bar{M})(\bar{Y}_i-\bar{\bar{Y}}_N)\\ &= \bar{\bar{Y}}_N-\bar{Y}\end{aligned}$$

where $\bar{\bar{y}}_n = \frac{1}{n}\sum_{i=1}^{n}\bar{Y}_i.$

An unbiased estimator of population mean $\bar{Y}$ is thus obtained as

$$\bar{y}_{S2}+\frac{N-1}{N\bar{M}}\frac{1}{N-1}\sum_{i=1}^{n}(M_i-\bar{m})(\bar{y}_{i(mi)}-\bar{y}_{S2}).$$

Note that the bias arises due to inequality of sizes of the first stage units and probability of selection of second stage units varies from one first stage to another.

Variance

$$\begin{aligned}Var(\bar{y}_{S2}) &= E\left[Var(\bar{y}_{S2}\mid n)\right]+Var\left[E(\bar{y}_{S2}\mid n)\right]\\ &= Var\left[\frac{1}{n}\sum_{i=1}^{n}\bar{y}_i\right]+E\left[\frac{1}{n^2}\sum_{i=1}^{n}Var(\bar{y}_{i(mi)}\mid i)\right]\\ &= \left(\frac{1}{n}-\frac{1}{N}\right)S_b^2+E\left[\frac{1}{n^2}\sum_{i=1}^{n}\left(\frac{1}{m_i}-\frac{1}{M_i}\right)S_i^2\right]\\ &= \left(\frac{1}{n}-\frac{1}{N}\right)S_b^2+\frac{1}{Nn}\sum_{i=1}^{N}\left(\frac{1}{m_i}-\frac{1}{M_i}\right)S_i^2\end{aligned}$$

where

$$S_b^2 = \frac{1}{N-1}\sum_{i=1}^{N}\left(\overline{Y}_i - \overline{\overline{Y}}_N\right)^2$$

$$S_i^2 = \frac{1}{M_i - 1}\sum_{j=1}^{M_i}\left(y_{ij} - \overline{Y}_i\right)^2.$$

The MSE can be obtained as

$$MSE(\overline{y}_{S2}) = Var(\overline{y}_{S2}) + \left[Bias(\overline{y}_{S2})\right]^2.$$

Estimation of Variance

Consider mean square between cluster means in the sample

$$s_b^2 = \frac{1}{n-1}\sum_{i=1}^{n}\left(\overline{y}_{i(mi)} - \overline{y}_{S2}\right)^2.$$

It can be shown that

$$E(s_b^2) = S_b^2 + \frac{1}{N}\sum_{i=1}^{N}\left(\frac{1}{m_i} - \frac{1}{M_i}\right)S_i^2$$

Also $s_i^2 = \dfrac{1}{m_i - 1}\sum_{j=1}^{m_i}(y_{ij} - \overline{y}_{i(mi)})^2$

$$E(s_i^2) = S_i^2 = \frac{1}{M_i - 1}\sum_{j=1}^{M_i}(y_{ij} - \overline{Y}_i)^2$$

So $E\left[\dfrac{1}{n}\sum_{i=1}^{n}\left(\dfrac{1}{m_i} - \dfrac{1}{M_i}\right)s_i^2\right] = \dfrac{1}{N}\sum_{i=1}^{N}\left(\dfrac{1}{m_i} - \dfrac{1}{M_i}\right)S_i^2.$

Thus

$$E(s_b^2) = S_b^2 + E\left[\frac{1}{n}\sum_{i=1}^{n}\left(\frac{1}{m_i} - \frac{1}{M_i}\right)s_i^2\right]$$

and an unbiased estimator of S_b^2 is

$$\hat{S}_b^2 = s_b^2 - \frac{1}{n}\sum_{i=1}^{n}\left(\frac{1}{m_i} - \frac{1}{M_i}\right)s_i^2.$$

So an estimator of variance can be obtained by replacing S_b^2 *and* S_i^2 by their unbiased estimators as

$$\widehat{Var}(\bar{y}_{S2}) = \left(\frac{1}{n} - \frac{1}{N}\right)\hat{S}_b^2 + \frac{1}{Nn}\sum_{i=1}^{N}\left(\frac{1}{m_i} - \frac{1}{M_i}\right)\hat{S}_i^2$$

2. Estimation based on first stage unit totals :

$$\hat{\bar{Y}} = \bar{y}_{S2}^* = \frac{1}{n}\sum_{i=1}^{n}\frac{M_i\bar{y}_{i(mi)}}{\bar{M}}$$

$$= \frac{1}{n}\sum_{i=1}^{n}u_i\bar{y}_{i(mi)}$$

where $u_i = \dfrac{M_i}{\bar{M}}.$

Bias

$$E(y_{S2}^*) = E\left[\frac{1}{n}\sum_{i=1}^{n}u_i\bar{y}_{i(mi)}\right]$$

$$= E\left[\frac{1}{n}\sum_{i=1}^{n}u_iE_2(\bar{y}_{i(mi)} \mid i)\right]$$

$$= E\left[\frac{1}{n}\sum_{i=1}^{n}u_i\bar{Y}_i\right]$$

$$= \frac{1}{N}\sum_{i=1}^{N}u_i\bar{Y}_i$$

$$= \bar{Y}.$$

Thus $\bar{y}_{S2}^*$ is an unbiased estimator of $\bar{Y}$.

Variance

$$Var(\bar{y}_{S2}^*) = Var\left[E(\bar{y}_{S2}^* \mid n)\right] + E\left[Var(\bar{y}_{S2}^* \mid n)\right]$$

$$= Var\left[\frac{1}{n}\sum_{i=1}^{n}u_i\bar{Y}_i\right] + E\left[\frac{1}{n^2}\sum_{i=1}^{n}u_i^2Var(\bar{y}_{i(mi)}) \mid i\right]$$

$$= \left(\frac{1}{n} - \frac{1}{N}\right)S_b^{*2} + \frac{1}{nN}\sum_{i=1}^{N}u_i^2\left(\frac{1}{m_i} - \frac{1}{M_i}\right)S_i^2$$

where

$$S_i^2 = \frac{1}{M_i - 1}\sum_{j=1}^{M_i}(y_{ij} - \overline{Y}_i)^2$$

$$S_b^{*2} = \frac{1}{N-1}\sum_{j=1}^{N}(u_i\overline{Y}_i - \overline{Y})^2.$$

3. Estimator based on ratio estimator:

$$\hat{\overline{Y}} = \overline{y}_{S2}^{**} = \frac{\sum_{i=1}^{n} M_i \overline{y}_{i(mi)}}{\sum_{i=1}^{n} M_i} = \frac{\sum_{i=1}^{n} u_i \overline{y}_{i(mi)}}{\sum_{i=1}^{n} u_i} = \frac{\overline{y}_{S2}^{*}}{\overline{u}_n}$$

where $u_i = \frac{M_i}{\overline{M}}$, $\overline{u}_n = \frac{1}{n}\sum_{i=1}^{n} u_i.$

This estimator can be seen as if arising by ratio method of estimation as follows:

Let

$$y_i^* = u_i \overline{y}_{i(mi)}$$

$$x_i^* = \frac{M_i}{\overline{M}},\ i = 1, 2, ..., N$$

be the values of study variable and auxiliary variable in reference to ratio method of estimation. Then

$$\overline{y}^* = \frac{1}{n}\sum_{i=1}^{n} y_i^* = \overline{y}_{S2}^*$$

$$\overline{x}^* = \frac{1}{n}\sum_{i=1}^{n} x_i^* = \overline{u}_n$$

$$\overline{X}^* = \frac{1}{N}\sum_{i=1}^{N} X_i^* = 1.$$

The corresponding ratio estimator of $\overline{Y}$ is

$$\hat{\overline{Y}}_R = \frac{\overline{y}^*}{\overline{x}^*}\overline{X}^* = \frac{\overline{y}_{S2}^*}{\overline{u}_n}1 = \overline{y}_{S2}^{**}.$$

So the bias and mean squared error of $\overline{y}_{S2}^{**}$ can be obtained directly from the results of ratio estimator. Recall that in ratio method of estimation, the bias of ratio estimator upto second order of approximation is

$$Bias(\hat{\overline{y}}_R) \approx \frac{N-n}{Nn}\overline{Y}(C_x^2 - 2\rho C_x C_y)$$

$$= \overline{Y}\left[\frac{Var(\overline{x})}{\overline{X}^2} - \frac{Cov(\overline{x},\overline{y})}{\overline{X}\overline{Y}}\right]$$

$$MSE(\hat{\overline{Y}}_R) \approx \left[Var(\overline{y}) + R^2 Var(\overline{x}) - 2RCov(\overline{x},\overline{y})\right]$$

where $R = \dfrac{\overline{Y}}{\overline{X}}$.

Bias

The bias of $\overline{y}_{S2}^{**}$ up to second order of approximation is

$$Bias(\overline{y}_{S2}^{**}) = \overline{Y}\left[\frac{Var(\overline{x}_{S2}^{*})}{\overline{X}^2} - \frac{Cov(\overline{x}_{S2}^{*}, \overline{y}_{S2}^{*})}{\overline{X}\overline{Y}}\right]$$

where $\overline{x}_{S2}^{*}$ is the mean of auxiliary variable similar to $\overline{y}_{S2}^{**}$ as $\overline{x}_{S2}^{*} = \dfrac{1}{n}\sum_{i=1}^{n}\overline{x}_{i(mi)}$.

Now we find $Cov(\overline{x}_{S2}^{*}, \overline{y}_{S2}^{*})$.

$$Cov(\overline{x}_{S2}^{*}, \overline{y}_{S2}^{*}) = Cov\left[E\left(\frac{1}{n}\sum_{i=1}^{n}u_i\overline{x}_{i(mi)}, \frac{1}{n}\sum_{i=1}^{n}u_i\overline{y}_{i(mi)}\right)\right] + E\left[Cov\left(\frac{1}{n}\sum_{i=1}^{n}u_i\overline{x}_{i(mi)}, \frac{1}{n}\sum_{i=1}^{n}u_i\overline{y}_{i(mi)}\right)\right]$$

$$= Cov\left[\frac{1}{n}\sum_{i=1}^{n}u_i E(\overline{x}_{i(mi)}), \frac{1}{n}\sum_{i=1}^{n}u_i E(\overline{y}_{i(mi)})\right] + E\left[\frac{1}{n^2}\sum_{i=1}^{n}u_i^2 Cov(\overline{x}_{i(mi)}, \overline{y}_{i(mi)}) \mid i\right]$$

$$= Cov\left[\frac{1}{n}\sum_{i=1}^{n}u_i\overline{X}_i, \frac{1}{n}\sum_{i=1}^{n}u_i\overline{Y}_i\right] + E\left[\frac{1}{n^2}\sum_{i=1}^{n}u_i^2\left(\frac{1}{m_i} - \frac{1}{M_i}\right)S_{ixy}\right]$$

$$= \left(\frac{1}{n} - \frac{1}{N}\right)S_{bxy}^{*} + \frac{1}{nN}\sum_{i=1}^{N}u_i^2\left(\frac{1}{m_i} - \frac{1}{M_i}\right)S_{ixy}$$

where

$$S_{bxy}^{*} = \frac{1}{N-1}\sum_{i=1}^{N}(u_i\overline{X}_i - \overline{X})(u_i\overline{Y}_i - \overline{Y})$$

$$S_{ixy} = \frac{1}{M_i - 1}\sum_{j=1}^{M_i}(x_{ij} - \overline{X}_i)(y_{ij} - \overline{Y}_i).$$

Similarly, $Var(\overline{x}_{S2}^{*})$ can be obtained by replacing x in place of y in $Cov(\overline{x}_{S2}^{*}, \overline{y}_{S2}^{*})$ as

$$Var(\overline{x}_{S2}^{*}) = \left(\frac{1}{n} - \frac{1}{N}\right)S_{bx}^{*2} + \frac{1}{nN}\sum_{i=1}^{N}u_i^2\left(\frac{1}{m_i} - \frac{1}{M_i}\right)S_{ix}^2$$

$$\text{where } S_{bx}^{*2} = \frac{1}{N-1}\sum_{i=1}^{N}(u_i\bar{X}_i - \bar{X})^2$$

$$S_{ix}^{*2} = \frac{1}{M_i-1}\sum_{i=1}^{M_i}(x_{ij} - \bar{X}_i)^2.$$

Substituting $Cov(\bar{x}_{S2}^{*}, \bar{y}_{S2}^{*})$ and $Var(\bar{x}_{S2}^{*})$ in $Bias(\bar{y}_{S2}^{**})$ we obtain the approximate bias as

$$Bias(\bar{y}_{S2}^{**}) \approx \bar{Y}\left[\left(\frac{1}{n}-\frac{1}{N}\right)\left(\frac{S_{bx}^{*2}}{\bar{X}^2}-\frac{S_{bxy}^{*}}{\bar{X}\bar{Y}}\right)+\frac{1}{nN}\sum_{i=1}^{N}\left\{u_i^2\left(\frac{1}{m_i}-\frac{1}{M_i}\right)\left(\frac{S_{ix}^2}{\bar{X}^2}-\frac{S_{ixy}}{\bar{X}\bar{Y}}\right)\right\}\right].$$

Mean Squared Error

$$MSE(\bar{y}_{S2}^{**}) \approx Var(\bar{y}_{S2}^{*}) - 2R^{*}Cov(\bar{x}_{S2}^{*}, \bar{y}_{S2}^{*}) + R^{*2}Var(\bar{x}_{S2}^{*})$$

$$Var(\bar{y}_{S2}^{**}) = \left(\frac{1}{n}-\frac{1}{N}\right)S_{by}^{*2} + \frac{1}{nN}\sum_{i=1}^{N}u_i^2\left(\frac{1}{m_i}-\frac{1}{M_i}\right)S_{iy}^2$$

$$Var(\bar{x}_{S2}^{**}) = \left(\frac{1}{n}-\frac{1}{N}\right)S_{bx}^{*2} + \frac{1}{nN}\sum_{i=1}^{N}u_i^2\left(\frac{1}{m_i}-\frac{1}{M_i}\right)S_{ix}^2$$

$$Cov(\bar{x}_{S2}^{*}, \bar{y}_{S2}^{**}) = \left(\frac{1}{n}-\frac{1}{N}\right)S_{bxy}^{*2} + \frac{1}{nN}\sum_{i=1}^{N}u_i^2\left(\frac{1}{m_i}-\frac{1}{M_i}\right)S_{ixy}^2$$

where

$$S_{by}^{*2} = \frac{1}{N-1}\sum_{i=1}^{N}(u_i\bar{Y}_i - \bar{Y})^2$$

$$S_{iy}^{*2} = \frac{1}{M_i-1}\sum_{j=1}^{M_i}(y_{ij} - \bar{Y}_i)^2$$

$$R^{*} = \frac{\bar{Y}}{\bar{X}} = \bar{Y}.$$

Thus

$$MSE(\bar{y}_{S2}^{**}) \approx \left(\frac{1}{n}-\frac{1}{N}\right)\left(S_{by}^{*2} - 2R^{*}S_{bxy}^{*} + R^{*2}S_{bx}^{*2}\right) + \frac{1}{nN}\sum_{i=1}^{N}\left[u_i^2\left(\frac{1}{m_i}-\frac{1}{M_i}\right)\left(S_{iy}^2 - 2R^{*}S_{ixy} + R^{*2}S_{ix}^2\right)\right].$$

Also

$$MSE(\bar{y}_{S2}^{**}) \approx \left(\frac{1}{n}-\frac{1}{N}\right)\frac{1}{N-1}\sum_{i=1}^{N}u_i^2\left(\bar{Y}_i - R^{*}\bar{X}_i\right)^2 + \frac{1}{nN}\sum_{i=1}^{N}\left[u_i^2\left(\frac{1}{m_i}-\frac{1}{M_i}\right)\left(S_{iy}^2 - 2R^{*}S_{ixy} + R^{*2}S_{ix}^2\right)\right].$$

Estimate of Variance

Consider

$$s_{bxy}^{*} = \frac{1}{n-1}\sum_{i=1}^{n}\left[\left(u_i\bar{y}_{i(mi)} - \bar{y}_{S2}^{*}\right)\left(u_i\bar{x}_{i(mi)} - \bar{x}_{S2}^{*}\right)\right]$$

$$s_{ixy} = \frac{1}{m_i - 1}\sum_{j=1}^{n}\left[\left(x_{ij} - \bar{x}_{i(mi)}\right)\left(y_{ij} - \bar{y}_{i(mi)}\right)\right].$$

It can be shown that

$$E\left(s^{*}_{bxy}\right) = S^{*}_{bxy} + \frac{1}{N}\sum_{i=1}^{N}u_i^2\left(\frac{1}{m_i} - \frac{1}{M_i}\right)S_{ixy}$$

$$E(s_{ixy}) = S_{ixy}.$$

So

$$E\left[\frac{1}{n}\sum_{i=1}^{n}u_i^2\left(\frac{1}{m_i} - \frac{1}{M_i}\right)s_{ixy}\right] = \frac{1}{N}\sum_{i=1}^{N}\left[u_i^2\left(\frac{1}{m_i} - \frac{1}{M_i}\right)S_{ixy}\right].$$

Thus

$$\hat{S}^{*}_{bxy} = s^{*}_{bxy} - \frac{1}{n}\sum_{i=1}^{n}u_i^2\left(\frac{1}{m_i} - \frac{1}{M_i}\right)s_{ixy}$$

$$\hat{S}^{*2}_{bx} = s^{*2}_{bx} - \frac{1}{n}\sum_{i=1}^{n}u_i^2\left(\frac{1}{m_i} - \frac{1}{M_i}\right)s^2_{ix}$$

$$\hat{S}^{*2}_{by} = s^{*2}_{by} - \frac{1}{n}\sum_{i=1}^{n}u_i^2\left(\frac{1}{m_i} - \frac{1}{M_i}\right)s^2_{iy}.$$

Also

$$E\left[\frac{1}{n}\sum_{i=1}^{n}\left\{u_i^2\left(\frac{1}{m_i} - \frac{1}{M_i}\right)s^2_{ix}\right\}\right] = \frac{1}{N}\sum_{i=1}^{N}\left[u_i^2\left(\frac{1}{m_i} - \frac{1}{M_i}\right)S^2_{ix}\right]$$

$$E\left[\frac{1}{n}\sum_{i=1}^{n}\left\{u_i^2\left(\frac{1}{m_i} - \frac{1}{M_i}\right)s^2_{iy}\right\}\right] = \frac{1}{N}\sum_{i=1}^{N}\left[u_i^2\left(\frac{1}{m_i} - \frac{1}{M_i}\right)S^2_{iy}\right].$$

A consistent estimator of MSE of $\bar{y}^{**}_{S2}$ can be obtained by substituting the unbiased estimators of respective statistics in $MSE(\bar{y}^{**}_{S2})$ as

$$\widehat{MSE}(\bar{y}^{**}_{S2}) \approx \left(\frac{1}{n} - \frac{1}{N}\right)\left(s^{*2}_{by} - 2r^{*}s^{*}_{bxy} + r^{*2}s^{*2}_{bx}\right) + \frac{1}{nN}\sum_{i=1}^{n}u_i^2\left(\frac{1}{m_i} - \frac{1}{M_i}\right)\left(s^2_{iy} - 2r^{*}s_{ixy} + r^{*2}s^2_{ix}\right)$$

$$\approx \left(\frac{1}{n} - \frac{1}{N}\right)\frac{1}{n-1}\sum_{i=1}^{n}\left(\bar{y}_{i(mi)} - r^{*}\bar{x}_{i(mi)}\right)^2 + \frac{1}{nN}\sum_{i=1}^{n}\left[u_i^2\left(\frac{1}{m_i} - \frac{1}{M_i}\right)\left(s^2_{iy} - 2r^{*}s_{ixy} + r^{*2}s^2_{ix}\right)\right]$$

where

$$r^{*} = \frac{\bar{y}^{*}_{S2}}{\bar{x}^{*}_{S2}}.$$

5

Systematic Sampling: A Comprehensive Study

Systematic sampling is a sampling technique that is methodical as the selection procedure follows a decided pattern. It is considered to be more convenient than random sampling method and the chances of equal probability of selection of elements remains same. The section serves as a source to understand the major categories related to systematic sampling. This chapter elucidates the crucial theories and principles of systematic sampling.

Systematic Sampling

Systematic sampling is a statistical method involving the selection of elements from an ordered sampling frame. The most common form of systematic sampling is an equiprobability method. In this approach, progression through the list is treated circularly, with a return to the top once the end of the list is passed. The sampling starts by selecting an element from the list at random and then every k^{th} element in the frame is selected, where k, the sampling interval (sometimes known as the *skip*): this is calculated as:

$$k = \frac{N}{n}$$

where n is the sample size, and N is the population size.

Using this procedure each element in the population has a known and equal probability of selection. This makes systematic sampling functionally similar to simple random sampling (SRS). However it is not the same as SRS because not every possible sample of a certain size has an equal chance of being chosen (e.g. samples with at least two elements adjacent to each other will never be chosen by systematic sampling). It is however, much more efficient (if variance within systematic sample is more than variance of population).

Systematic sampling is to be applied only if the given population is logically homogeneous, because systematic sample units are uniformly distributed over the population. The researcher must ensure that the chosen sampling interval does not hide a pattern. Any pattern would threaten randomness.

Example: Suppose a supermarket wants to study buying habits of their customers, then using systematic sampling they can choose every 10th or 15th customer entering the supermarket and conduct the study on this sample.

This is random sampling with a system. From the sampling frame, a starting point is chosen at random, and choices thereafter are at regular intervals. For example, suppose you want to sample 8 houses from a street of 120 houses. 120/8=15, so every 15th house is chosen after a random starting point between 1 and 15. If the random starting point is 11, then the houses selected are 11, 26, 41, 56, 71, 86, 101, and 116. As an aside, if every 15th house was a "corner house" then this corner pattern could destroy the randomness of the population.

If, as more frequently, the population is not evenly divisible (suppose you want to sample 8 houses out of 125, where 125/8=15.625), should you take every 15th house or every 16th house? If you take every 16th house, 8*16=128, so there is a risk that the last house chosen does not exist. On the other hand, if you take every 15th house, 8*15=120, so the last five houses will never be selected. The random starting point should instead be selected as a noninteger between 0 and 15.625 (inclusive on one endpoint only) to ensure that every house has equal chance of being selected; the interval should now be nonintegral (15.625); and each noninteger selected should be rounded up to the next integer. If the random starting point is 3.6, then the houses selected are 4, 20, 35, 50, 66, 82, 98, and 113, where there are 3 cyclic intervals of 15 and 4 intervals of 16.

To illustrate the danger of systematic skip concealing a pattern, suppose we were to sample a planned neighbourhood where each street has ten houses on each block. This places houses No. 1, 10, 11, 20, 21, 30... on block corners; corner blocks may be less valuable, since more of their area is taken up by streetfront etc. that is unavailable for building purposes. If we then sample every 10th household, our sample will either be made up *only* of corner houses (if we start at 1 or 10) or have *no* corner houses (any other start); either way, it will not be representative.

Systematic sampling may also be used with non-equal selection probabilities. In this case, rather than simply counting through elements of the population and selecting every k^{th} unit, we allocate each element a space along a number line according to its selection probability. We then generate a random start from a uniform distribution between 0 and 1, and move along the number line in steps of 1.

Example: We have a population of 5 units (A to E). We want to give unit A a 20% probability of selection, unit B a 40% probability, and so on up to unit E (100%). Assuming we maintain alphabetical order, we allocate each unit to the following interval:

A: 0 to 0.2

B: 0.2 to 0.6 (= 0.2 + 0.4)

C: 0.6 to 1.2 (= 0.6 + 0.6)

D: 1.2 to 2.0 (= 1.2 + 0.8)

E: 2.0 to 3.0 (= 2.0 + 1.0)

If our random start was 0.156, we would first select the unit whose interval contains this number (i.e. A). Next, we would select the interval containing 1.156 (element C), then 2.156 (element E). If instead our random start was 0.350, we would select from points 0.350 (B), 1.350 (D), and 2.350 (E).

Systematic sampling technique is operationally more convenient than simple random sampling. It also ensures at the same time that each unit has equal probability of inclusion in the sample. In this method of sampling, the first unit is selected with the help of random numbers and the remaining units are selected automatically according to a predetermined pattern. This method is known as systematic sampling.

Suppose the units in the population are numbered 1 to N in some order. Suppose further that N is expressible as a product of two integers n and k, so that $N = nk$.

To draw a sample of size n,

- select a random number between 1 and k.
- Suppose it is i.
- Select the first unit whose serial number is i.
- Select every k^{th} unit after i^{th} unit.
- Sample will contain $i, i + k, 1 + 2k, \ldots, i + (n - 1)k$ serial number units.

So first unit is selected at random and other units are selected systematically. This systematic sample is called kth systematic sample and k is termed as sampling interval. This is also known as linear systematic sampling.

The observations in systematic sampling are arranged as in the following table:

Systematic Sample Number		1	2	3	...	i	...	k
Sample composition	1	y_1	y_2	y_3	...	y_i	...	y_k
	2	y_{k+1}	y_{k+2}	y_{k+3}	...	y_{k+i}	...	y_{2k}
	⋮	⋮	⋮	⋮	⋮	⋮	⋮	⋮
	n	$y_{(n-1)k+1}$	$y_{(n-1)k+2}$	$y_{(n-1)k+3}$	...	$y_{(n-1)k+i}$	...	y_{nk}
Probability		$\frac{1}{k}$	$\frac{1}{k}$	$\frac{1}{k}$	...	$\frac{1}{k}$	...	$\frac{1}{k}$
Sample mean		$\bar{y}_1$	$\bar{y}_2$	$\bar{y}_3$	...	$\bar{y}_i$	...	$\bar{y}_k$

Example: Let $N = 50$ and $n = 5$. So $k = 10$. Suppose first selected number between 1 and 10 is 3. Then systematic sample consists of units with following serial number 3, 13, 23, 33, 43.

Systematic Sampling is Two Dimensions

Assume that the units in a population are arranged in the form of ml rows and each row contains nk units. A sample of size mn is required. Then

- select a pair of random numbers (i, j) such that $i \leq \ell$ and $j \leq k$.
- Select the $(i, j)^{th}$ unit, i.e., j^{th} unit in i^{th} row as the first unit.
- Then the rows to be selected

  $i, i + \ell, i + 2\ell, \ldots, i + (m - 1)\ell$

  and columns to be selected are

  $j, j + k, j + 2k, \ldots, j + (n - 1)k.$
- The points at which the m selected rows and n selected columns intersect determine the position of mn selected units in the sample.

Such a sample is called an aligned sample.

Alternative approach to select the sample is

- independently select n random integers $i_1, i_2, \ldots, i_n$ such that each of them is less than or equal to ℓ.
- Independently select m random integers $j_1, j_2, \ldots, j_m$ such that each of them is less than or equal to k.
- The units selected in the sample will have following coordinates.

$$(i_1 + r\ell, j_{r+1}), (i_2 + r\ell, j_{r+1} + k), (i_3 + r\ell, j_{r+1} + 2k), \ldots, (i_n + r\ell, j_{r+1} + (n-1)k)$$

Such a sample is called an unaligned sample.

Under certain conditions, an unaligned sample is often superior to an aligned sample as well as a stratified random sample.

Advantages of Systematic Sampling

1. It is easier to draw a sample and often easier to execute it without mistakes. This is more advantageous when the drawing is done in fields and offices as there may be substantial saving in time.

2. The cost is low and selection of units is simple. Much less training is needed for surveyors to collect units through systematic sampling.

3. The systematic sample is spread more evenly over the population. So no large part will fail to be represented in the sample. The sample is evenly spread and cross section is better. Systematic sampling fails in case of too many blanks.

Relation to Cluster Sampling

The systematic sample can be viewed from the cluster sampling point of view. With n = nk, there are k possible systematic samples. The same population can be viewed as if divided into k large sampling units, each of which contains n of the original units. The operation of choosing a systematic sample is equivalent to choosing one of the large sampling unit at random which constitutes the whole sample. A systematic sample is thus a simple random sample of one cluster unit from a population of k cluster units.

Estimation of Population Mean : When N = nk

Let

y_{ij} : observation on the unit bearing the serial number $i+(j-1)k$ in population, $i=1,2,...k,\ j=1,2,...n.$

Suppose the drawn random number is $i \le k.$

Sample consists of ith column (in earlier table).

Consider

$$\bar{y}_{sy} = \bar{y}_i = \frac{1}{n}\sum_{j=1}^{n} y_{ij}$$

sample mean as an estimator of population mean given by

$$\bar{Y} = \frac{1}{nk}\sum_{i=1}^{k}\sum_{j=1}^{n} y_{ij}$$

$$= \frac{1}{nk}\sum_{i=1}^{k} \bar{y}_i$$

Probability of selecting i^{th} column as systematic sample $= \frac{1}{k}$

So

$$E(\bar{y}_{sy}) = \frac{1}{k}\sum_{i=1}^{k} \bar{y}_i = \bar{Y}.$$

Thus $\overline{y}_{sy}$ is an unbiased estimator of $\overline{Y}$.

Further,

$$Var(\overline{y}_{sy}) = \frac{1}{k}\sum_{i=1}^{k}(\overline{y}_i - \overline{Y})^2.$$

Consider

$$\begin{aligned}(N-1)S^2 &= \sum_{i=1}^{k}\sum_{j=1}^{n}(y_{ij} - \overline{Y})^2 \\ &= \sum_{i=1}^{k}\sum_{j=1}^{n}\left[(y_{ij} - \overline{y}_i) + (\overline{y}_i - \overline{Y})\right]^2 \\ &= \sum_{i=1}^{k}\sum_{j=1}^{n}(y_{ij} - \overline{y}_i)^2 + n\sum_{i=1}^{k}(\overline{y}_i - \overline{Y})^2 \\ &= k(n-1)S_{wsy}^2 + n\sum_{i=1}^{k}(\overline{y}_i - \overline{Y})^2\end{aligned}$$

where

$$S_{wsy}^2 = \frac{1}{k(n-1)}\sum_{i=1}^{k}\sum_{j=1}^{n}(y_{ij} - \overline{y}_i)^2$$

is the variation among units that lie within the same systematic sample. Thus

$$Var(\overline{y}_{sy}) = \frac{N-1}{N}S^2 - \frac{k(n-1)}{N}S_{wsy}^2$$

$$= \frac{N-1}{N}S^2 - \frac{(n-1)}{n}S_{wsy}^2$$

Variation as a whole	Pooled within variation of the k systematic sample

with N=nk. This expression indicates that when the within variation is large, then $Var(\overline{y}_i)$ becomes smaller. Thus higher heterogeneity makes the estimator more efficient and higher heterogeneity in well expected is systematic sample.

Alternative form of Variance

$$Var(\overline{y}_{sy}) = \frac{1}{k}\sum_{i=1}^{k}(\overline{y}_i - \overline{Y})^2$$

$$= \frac{1}{k}\sum_{i=1}^{k}\left[\frac{1}{n}\sum_{j=1}^{n} y_{ij} - \bar{Y}\right]^2 = \frac{1}{kn^2}\sum_{i=1}^{k}\left[\sum_{j=1}^{n}(y_{ij} - \bar{Y})\right]^2$$

$$= \frac{1}{kn^2}\sum_{i=1}^{k}\left[\sum_{j=1}^{n}(y_{ij} - \bar{Y})^2 + \sum_{j(\neq \ell)=1}^{n}\sum_{\ell=1}^{n}(y_{ij} - \bar{Y})(y_{i\ell} - \bar{Y})\right]$$

$$= \frac{1}{kn^2}\left[(nk-1)S^2 + \sum_{i=1}^{k}\sum_{j(\neq \ell)=1}^{n}\sum_{\ell=1}^{n}(y_{ij} - \bar{Y})(y_{i\ell} - \bar{Y})\right].$$

The intraclass correlation between the pairs of units that are in the same systematic sample is

$$\rho_w = \frac{E(y_{ij} - \bar{Y})(y_{i\ell} - \bar{Y})}{E(y_{ij} - \bar{Y})^2}; \qquad -\frac{1}{nk-1} \le \rho \le 1$$

$$= \frac{\frac{1}{nk(n-1)}\sum_{i=1}^{k}\sum_{j(\neq \ell)=1}^{n}\sum_{\ell=1}^{n}(y_{ij} - \bar{Y})(y_{i\ell} - \bar{Y})}{\left(\frac{nk-1}{nk}\right)S^2}.$$

So substituting

$$\sum_{i=1}^{k}\sum_{j(\neq \ell)=1}^{n}\sum_{\ell=1}^{n}(y_{ij} - \bar{Y})(y_{i\ell} - \bar{Y}) = (n-1)(nk-1)\rho_w S^2$$

$Var(\bar{y}_i)$ gives $Var(\bar{y}_{sy}) = \frac{nk-1}{nk}\frac{S^2}{n}\left[1 + \rho_w(n-1)\right]$

$$= \frac{N-1}{N}\frac{S^2}{n}\left[1 + \rho_w(n-1)\right].$$

Comparison with SRSWOR

For a SRSWOR sample of size n,

$$Var(\bar{y}_{SRS}) = \frac{N-n}{Nn}S^2$$

$$= \frac{nk-n}{Nn}S^2$$

$$= \frac{k-1}{N}S^2.$$

Since

$$Var(\overline{y}_{sy}) = \frac{N-1}{N}S^2 - \frac{n-1}{n}S_{wsy}^2$$

$$N = nk$$

$$Var(\overline{y}_{SRS}) - Var(\overline{y}_{sy}) = \left(\frac{k-1}{N} - \frac{N-1}{N}\right)S^2 + \frac{n-1}{n}S_{wsy}^2$$

$$= \frac{n-1}{n}(S_{wsy}^2 - S^2).$$

Thus $\overline{y}_{sy}$ is

- more efficient than $\overline{y}_{SRS}$ when $S_{wsy}^2 > S^2$.
- less efficient than $\overline{y}_{SRS}$ when $S_{wsy}^2 < S^2$.
- equally efficient as $\overline{y}_{SRS}$ when $S_{wsy}^2 = S^2$.

Also, the relative efficiency of $\overline{y}_{sy}$ relative to $\overline{y}_{SRS}$ is

$$RE = \frac{Var(\overline{y}_{SRS})}{Var(\overline{y}_{sy})}$$

$$= \frac{\frac{N-n}{Nn}S^2}{\frac{N-1}{Nn}S^2[1+\rho_w(n-1)]}$$

$$= \frac{N-n}{N-1}\left[\frac{1}{1+\rho_w(n-1)}\right]$$

$$= \frac{n(k-1)}{(nk-1)}\left[\frac{1}{1+\rho_w(n-1)}\right]; \qquad -\frac{1}{nk-1} \le \rho \le 1.$$

Thus $\overline{y}_{sy}$ is

- more efficient than $\overline{y}_{SRS}$ when $\rho_w < -\frac{1}{nk-1}$
- less efficient than $\overline{y}_{SRS}$ when $\rho_w > -\frac{1}{nk-1}$
- equally efficient as $\overline{y}_{SRS}$ when $\rho_w = -\frac{1}{nk-1}$.

Comparison with Stratified Sampling

The systematic sample can also be viewed as if arising as a stratified sample. If population of $N = nk$ units is divided into n strata and suppose one unit is randomly drawn from each one of the strata then we get a stratified sample of size n. In doing so, just consider each row of the following arrangement as a stratum.

Recall that in case of stratified sampling with k strata, the stratum mean

$$\overline{y}_{st} = \frac{1}{N}\sum_{j=1}^{k} N_j \overline{y}_j$$

is an unbiased estimator of population mean.

Considering the set up of stratified sample in the set up of systematic sample, we have

- Number of strata = n
- Size of strata = k (row size)
- Sample size to be drawn from each stratum = 1

And $\overline{y}_{st}$ becomes

$$\overline{y}_{st} = \frac{1}{nk}\sum_{j=1}^{n} k\overline{y}_j$$

$$= \frac{1}{n}\sum_{j=1}^{n} \overline{y}_j$$

$$Var(\overline{y}_{st}) = \frac{1}{n^2}\sum_{j=1}^{n} Var(\overline{y}_j)$$

$$= \frac{1}{n^2}\sum_{j=1}^{n} \frac{k-1}{k.1} S_j^2 \quad \left(\text{using } Var(\overline{y}_{SRS}) = \frac{N-n}{Nn} S^2\right)$$

$$= \frac{k-1}{kn^2}\sum_{j=1}^{n} S_j^2$$

$$= \frac{k-1}{nk} S_{wst}^2$$

$$= \frac{N-n}{Nn} S_{wst}^2$$

where

$$S_j^2 = \frac{1}{k-1}\sum_{i=1}^{k} (y_{ij} - \overline{y}_j)^2$$

is the mean sum of squares of units in j[th] stratum.

$$S_{wst}^2 = \frac{1}{n}\sum_{j=1}^{n} S_j^2$$

$$= \frac{1}{n(k-1)}\sum_{i=1}^{k}\sum_{j=1}^{n}(y_{ij} - \bar{y}_j)^2$$

is the mean sum of squares within strata (or rows).

The variance of systematic sample mean is

$$Var(\bar{y}_{sy}) = \frac{1}{k}\sum_{i=1}^{k}(\bar{y}_i - \bar{Y})^2$$

$$= \frac{1}{k}\sum_{i=1}^{k}\left[\frac{1}{n}\sum_{j=1}^{n} y_{ij} - \frac{1}{n}\sum_{j=1}^{n}\bar{y}_j\right]^2$$

$$= \frac{1}{n^2 k}\sum_{i=1}^{k}\left[\sum_{j=1}^{n}(y_{ij} - \bar{y}_j)\right]^2$$

$$= \frac{1}{n^2 k}\left[\sum_{i=1}^{k}\sum_{j=1}^{n}(y_{ij} - \bar{y}_j)^2 + \sum_{i=1}^{k}\sum_{j\neq\ell=1}^{n}(y_{ij} - \bar{y}_j)(y_{i\ell} - \bar{y}_\ell)\right].$$

Now we simplify and express this expression in terms of intraclass correlation coefficient. The intraclass correlation coefficient between pairs of deviations of units which lie along the same row measured from their stratum means is defined as

$$\rho_{wst} = \frac{\dfrac{1}{nk(n-1)}\displaystyle\sum_{i=1}^{k}\sum_{j\neq\ell=1}^{n}(y_{ij} - \bar{y}_j)(y_{i\ell} - \bar{y}_\ell)}{\dfrac{1}{nk}\displaystyle\sum_{i=1}^{k}\sum_{j=1}^{n}(y_{ij} - \bar{y}_j)^2}$$

$$= \frac{\displaystyle\sum_{i=1}^{k}\sum_{j(\neq\ell)=1}^{n}(y_{ij} - \bar{y}_j)(y_{i\ell} - \bar{y}_\ell)}{(N-n)(n-1)S_{wst}^2}$$

So

$$Var(\bar{y}_{sy}) = \frac{1}{n^2 k}\left[(N-n)S_{wst}^2 + (N-n)(n-1)\rho_{wst}S_{wst}^2\right]$$

$$= \frac{N-n}{Nn}S_{wst}^2\left[1 + (n-1)\rho_{wst}\right] \quad \text{(using } N = nk\text{)}.$$

Thus

$$Var(\bar{y}_{st}) - Var(\bar{y}_{sy}) = -\frac{N-n}{Nn}(n-1)\rho_{wst}S_{wst}^2$$

and the relative efficiency of systematic sampling relative to equivalent stratified sampling is given by

$$RE = \frac{1}{1+(n-1)\rho_{wst}}.$$

So the systematic sampling is

- more efficient than the corresponding equivalent stratified sample when $\rho_{wst} > 0$
- less efficient than the corresponding equivalent stratified sample when $\rho_{wst} < 0$
- equally efficient than the corresponding equivalent stratified sample when $\rho_{wst} = 0$

Comparison of Systematic Sampling, Stratified sampling and SRS with Population with Linear Trend

We assume that the values of units in the population increase according to linear trend.

So the values of successive units in the population increase in accordance with a linear model so that

$$y_i = a + bi, i = 1,2,...,N.$$

Now we determine the variances of $\bar{y}_{SRS}, \bar{y}_{sy}$ *and* $\bar{y}_{st}$ under this linear trend.

Under SRSWOR

$$V(\bar{y}_{SRS}) = \frac{N-n}{Nn}S^2$$

Here

$$N = nk$$

$$\bar{Y} = a + b\frac{1}{N}\sum_{i=1}^{N} i$$

$$= a + b\frac{1}{N}\frac{N(N+1)}{2} = a + b\frac{N+1}{2}$$

$$S^2 = \frac{1}{N-1}\sum_{i=1}^{N}(y_i - \bar{Y})^2$$

$$= \frac{1}{N-1} \sum_{i=1}^{N} \left[a + bi - a - b\frac{N+1}{2} \right]^2$$

$$= \frac{b^2}{N-1} \sum_{i=1}^{N} \left(i - \frac{N+1}{2} \right)^2$$

$$= \frac{b^2}{N-1} \left[\sum_{i=1}^{N} i^2 - N\left(\frac{N+1}{2} \right)^2 \right]$$

$$= \frac{b^2}{N-1} \left[\frac{N(N+1)(2N+1)}{6} - \frac{N(N+1)^2}{4} \right]$$

$$= b^2 \frac{N(N+1)}{12}$$

$$Var(\overline{y}_{SRS}) = \frac{nk-n}{nk.n} b^2 \frac{nk(nk+1)}{12}$$

$$= \frac{b^2}{12}(k-1)(nk+1).$$

Under Systematic Sampling

Earlier y_{ij} denoted the value of study variable with the jth unit in the ith systematic sample. Now y_{ij} represents the value of $[i + (j - 1)k]^{th}$ unit of the population, so

$$y_{ij} = a + b\left[i + (j-1)k\right], \; i = 1,2,\ldots,k; j = 1,2,\ldots,n.$$

$$\overline{y}_{sy} = \frac{1}{k}\sum_{i=1}^{k} \overline{y}_i$$

$$Var(\overline{y}_{sy}) = \frac{1}{k}\sum_{i=1}^{k} (\overline{y}_i - \overline{Y})^2$$

$$\overline{y}_i = \frac{1}{n}\sum_{j=1}^{n} y_{ij} \quad = \quad \frac{1}{n}\sum_{j=1}^{n} \left[a + b\{i + (j-1)k\}\right]$$

$$= a + b\left(i + \frac{n-1}{2}k \right)$$

$$\sum_{i=1}^{k} (\overline{y}_i - \overline{Y})^2 = \sum_{i=1}^{k} \left[a + b\left(i + \frac{n-1}{2}k \right) - a - b\frac{nk+1}{2} \right]^2$$

$$= b^2 \sum_{i=1}^{k} \left(i - \frac{k+1}{2} \right)^2 \quad = \quad b^2 \left[\sum_{i=1}^{k} i^2 + \left(\frac{k+1}{2} \right)^2 - 2\frac{k+1}{2}\sum_{i=1}^{k} i \right]$$

$$= b^2\left[\frac{k(k+1)(2k+1)}{6} + \left(\frac{k+1}{2}\right)^2 - (k+1)\frac{k(k+1)}{2}\right]$$

$$= \frac{b^2}{12}k(k^2-1)$$

$$Var(\overline{y}_{sy}) = \frac{1}{k}\frac{b^2}{12}k(k^2-1) \quad = \quad \frac{b^2}{12}(k^2-1).$$

Under Stratified Sampling

$$y_{ij} = a + b\left[i + (j-1)k\right],\ i = 1,2,\ldots,k,\ j = 1,2,\ldots,n$$

$$\overline{y}_{st} = \frac{1}{N}\sum_{i=1}^{k} N_i \overline{y}_i$$

$$Var(\overline{y}_{st}) = \frac{N-n}{Nn}S_{wst}^2$$

$$= \frac{k-1}{nk}S_{wst}^2$$

where $S_{wst}^2 = \frac{1}{n}\sum_{j=1}^{n} S_j^2$

$$= \frac{1}{n(k-1)}\sum_{i=1}^{k}\sum_{j=1}^{n}(y_{ij} - \overline{y}_j)^2$$

$$= \frac{1}{n(k-1)}\sum_{i=1}^{k}\sum_{j=1}^{n}\left[a + b\{i + (j-1)k\} - a - b\left\{\frac{k+1}{2} + (j-1)k\right\}\right]^2$$

$$= \frac{b^2}{n(k-1)}\sum_{i=1}^{k}\sum_{j=1}^{n}\left(i - \frac{k+1}{2}\right)^2$$

$$= \frac{b^2}{n(k-1)}\frac{nk(k^2-1)}{12}$$

$$= b^2\frac{k(k+1)}{12}$$

$$Var(\overline{y}_{st}) = \frac{k-1}{nk}b^2\frac{k(k+1)}{12}$$

$$= \frac{b^2}{12}\left(\frac{k^2-1}{n}\right).$$

If k is large, so $\frac{1}{k}$ that is negligible, then comparing $Var(\overline{y}_{st}), Var(\overline{y}_{sy})\, and\, V(\overline{y}_{SRS})$,

	$Var(\bar{y}_{st})$	:	$Var(\bar{y}_{sy})$	:	$Var(\bar{y}_{SRS})$
or	$\frac{k^2-1}{n}$	:	k^2-1	:	$(k-1)(1+nk)$
or	$\frac{k+1}{n}$	:	$k+1$	:	$nk+1$
or	$\frac{k+1}{n(k+1)}$	:	$\frac{k+1}{k+1}$	:	$\frac{nk+1}{k+1}$
$\simeq$	$\frac{1}{n}$		1	:	n

Thus

$$Var(\bar{y}_{st}) \;:\; Var(\bar{y}_{sy}) \;:\; V(\bar{y}_{SRS}) \;::\; \frac{1}{n} \;:\; 1 \;:\; n$$

So stratified sampling is best for linearly trended population. Next best is systematic sampling.

Estimation of Variance

As such there is only one cluster, so variance in principle, cannot be estimated.

Some approximations have been suggested.

1. Treat systematic sample as if it were a random sample.

$$\widehat{Var}(\bar{y}_{sy}) = \left(\frac{1}{n} - \frac{1}{nk}\right) s_{wc}^2$$

$$\text{where} \quad s_{wc}^2 = \frac{1}{n-1}\sum_{j=0}^{n-1}(y_{i+jk} - \bar{y}_i)^2.$$

This estimator under-estimates the true variance.

2. Use of successive differences of the values gives

$$\widehat{Var}(\bar{y}_{sy}) = \left(\frac{1}{n} - \frac{1}{nk}\right)\frac{1}{2(n-1)}\sum_{j=0}^{n-1}\left(y_{i+jk} - y_{i+(j+1)k}\right)^2.$$

This estimator is a biased estimator of true variance.

3. Use the balanced difference of $y_1, y_2, y_3, \ldots\ldots y_n$ to get

$$\widehat{Var}(\bar{y}_{sy}) = \left(\frac{1}{n} - \frac{1}{nk}\right)\frac{1}{5(n-2)}\sum_{i}^{n-2}\left[\frac{y_i}{2} - y_{i+1} + \frac{y_{i+2}}{2}\right]^2$$

or

$$\widehat{Var}(\bar{y}_{sy}) = \left(\frac{1}{n} - \frac{1}{nk}\right)\frac{1}{15(n-4)}\sum_{i}^{n-4}\left[\frac{y_i}{2} - y_{i+1} + y_{i+2} - y_{i+3} + \frac{y_{i+4}}{2}\right]^2.$$

4. The interpenetrating subsamples can be utilized by dividing the sample into C groups each of size $\frac{n}{c}$. Then the group means are $\bar{y}_1, \bar{y}_2,, ...\bar{y}_c$. Now find

$$\bar{y} = \frac{1}{c}\sum_{t=1}^{c}\bar{y}_t$$

$$\widehat{Var}(\bar{y}_{sy}) = \frac{1}{c(c-1)}\sum_{t=1}^{c}(y_t - \bar{y})^2.$$

Systematic sampling when $N \neq nk$.

When N is not expressible as nk then suppose N can be expressed as

$$N = nk + p;\ p < k.$$

Then consider the following sample mean as an estimator of population mean

$$\bar{y}_{sy} = \bar{y}_i = \begin{cases} \frac{1}{n+1}\sum_{j=1}^{n+1} y_{ij} & \text{if } i \leq p \\ \frac{1}{n}\sum_{j=1}^{n} y_{ij} & \text{if } i > p. \end{cases}$$

In this case

$$E(\bar{y}_i) = \frac{1}{k}\left[\sum_{i=1}^{p}\left(\frac{1}{n+1}\sum_{j=1}^{n+1} y_{ij}\right) + \sum_{i=p+1}^{n}\left(\frac{1}{n}\sum_{j=1}^{n} y_{ij}\right)\right]$$
$$\neq \bar{Y}.$$

So $\bar{y}_{sy}$ is a biased estimator of $\bar{Y}$.

An unbiased estimator of $\bar{Y}$ is

$$\bar{y}_{sy}^* = \frac{k}{N}\sum_{j} y_{ij}$$
$$= \frac{k}{N}C_i$$

where $C_i = n\bar{y}_i$ is the total of values of the ith column.

$$E(\bar{y}_{sy}^*) = \frac{k}{N}E(C_i)$$
$$= \frac{k}{N}\cdot\frac{1}{k}\sum_{i=1}^{k} C_i$$
$$= \bar{Y}$$

$$Var(\overline{y}_{sy}^{*}) = \frac{k^2}{N^2}\left(\frac{k-1}{k}\right)S_c^{*2}$$

where

$$S_c^{*2} = \frac{1}{k-1}\sum_{i=1}^{k}\left(n\overline{y}_i - \frac{N\overline{Y}}{k}\right)^2.$$

Now we consider another procedure which is opted when population size N is not expressible as the product of n and k.

When population size N is not expressible as the product of n and k, then let

$$N = nq + r.$$

Then take the sampling interval as

$$k = \begin{cases} q & if\ r \le \frac{n}{2} \\ q+1 & if\ r > \frac{n}{2} \end{cases}.$$

Let $\left[\frac{M}{g}\right]$ denotes the largest integer contained in $\frac{M}{g}$.

If $k = q^{*} (= q\ or\ q+1)$, then the number of units expected in sample =

$$\begin{cases} \left[\frac{N}{q^*}\right] & \text{with probability } \left[\frac{N}{q^*}\right] + 1 - \left(\frac{N}{q^*}\right) \\ \left[\frac{N}{q^*}\right] + 1 & \text{with probability } \left(\frac{N}{q^*}\right) - \left[\frac{N}{q^*}\right]. \end{cases}$$

If q = q*, then we get

$$n^{*} = \begin{cases} n + \left[\frac{r}{q}\right] \text{with probability } \left(\frac{r}{q}\right) + 1 - \left[\frac{r}{q}\right] \\ n + \left[\frac{r}{q}\right] + 1 \text{with probability } \left(\frac{r}{q}\right) - \left[\frac{r}{q}\right] \end{cases}$$

Similarly, if q* = q + 1, then

$$n^{*} = \begin{cases} n - \left(\frac{n-r}{q+1}\right) \text{with probability } \left[\frac{(n-r)}{(q+1)}\right] + 1 - \left(\frac{n-r}{q+1}\right) \\ n + \left[\left(\frac{n-r}{q+1}\right) + 1\right] \text{with probability } \left(\frac{n-r}{q+1}\right) - \left[\frac{(n-r)}{(q+1)}\right]. \end{cases}$$

Example: Let N =17 and n = 5. Then q = 3 and r = 2.Since $r<\frac{n}{2}, k=q=3.$

Then sample sizes would be

$$n^{*}=\begin{cases} n+\left[\frac{r}{q}\right]=5 \text{ with probability } \left[\frac{r}{q}\right]+1-\left(\frac{r}{q}\right)=\frac{1}{3} \\ n+\left[\frac{r}{q}\right]+1=6 \text{ with probability } \left(\frac{r}{q}\right)-\left[\frac{r}{q}\right]=\frac{2}{3}. \end{cases}$$

This can be verified from the following example:

Systematic sample number	Systematic sample	Probability
1	$Y_1, Y_4, Y_7, Y_{10}, Y_{13}, Y_{16}$	1 / 3
2	$Y_4, Y_5, Y_8, Y_{11}, Y_{14}, Y_{17}$	1 / 3
3	$Y_3, Y_6, Y_9, Y_{12}, Y_{15}$	1 / 3

We now prove the following theorem which shows how to obtain an unbiased estimator of the population mean when $N \neq nk.$

Theorem: In systematic sampling with sampling interval k from a population with size $N \neq nk$ an unbiased estimator of the population mean $\bar{Y}$ is given by

$$\hat{\bar{Y}}=\frac{k}{N}\left(\sum^{n'} y\right)_i$$

where i stands for the ith systematic sample $i=1,2,....,k$, and n' denotes the size of ith systematic sample.

Proof. Each systematic sample has probability $\frac{1}{k}$. Hence

$$E(\hat{\bar{Y}})=\sum_{i=1}^{k}\frac{1}{k}\cdot\frac{k}{N}\left(\sum^{n'} y\right)_i$$

$$=\frac{1}{N}\sum_{i=1}^{k}\left(\sum^{n'} y\right)_i.$$

Now, each unit occurs in only one of the k possible systematic samples. Hence

$$\sum_{i=1}^{k}\left(\sum^{n'} y\right)_i=\sum_{i=1}^{N} Y_i,$$

which on substitution in $E(\hat{\bar{Y}})$ proves the theorem.

When $N \neq nk$ the systematic samples are not of the same size and the sample mean is not an unbiased estimator of the population mean. To overcome these disadvantages of systematic sampling when $N \neq nk$, circular systematic sampling is proposed. Circular systematic sampling consists of selecting a random number from 1 to N and selecting, the unit corresponding to this random number and thereafter every kth unit in a cyclical manner till a sample of n units is obtained, k being the nearest integer to N/n. In other words, if i is a number selected at random from 1 to N, then the circular systematic sample consists of units with serial numbers

$$\left.\begin{array}{ll} i+jk, & \text{if } i=jk \leq N \\ i+jk-N, & \text{if } i=jk>N \end{array}\right\} \qquad j=0,1,2,\ldots,(n-1).$$

This sampling scheme ensures equal probability of inclusion in the sample for every unit.

Example

Let N = 14 and n = 5. Then, k = nearest integer to $\frac{14}{5}=3$. Let the first number selected at random from 1 to 14 be 7. Then, the circular systematic sample consists of units with serial numbers

7, 10, 13, 16 - 14=2, 19 −14 = 5.

This procedure is illustrated diagrammatically in following figure.

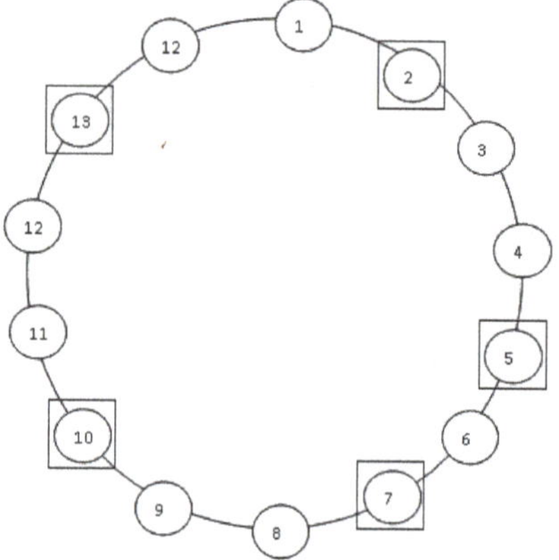

Theorem

In circular systematic sampling, the sample mean is an unbiased estimator of the population mean.

Proof: If i is the number selected at random, then the circular systematic sample mean is

$$\bar{y} = \frac{1}{n}\left(\sum^{n} y\right)_i,$$

where $\left(\sum^{n} y\right)_i$ denotes the total of y values in the ith circular systematic sample, i =1,2,...N. We note here that in circular systematic sampling, there are N circular systematic samples, each having probability $\frac{1}{N}$ of its selection.

Hence,

$$E(\bar{y}) = \sum_{i=1}^{N} \frac{1}{n}\left(\sum^{n} y\right)_i \times \frac{1}{N} = \frac{1}{Nn}\sum_{i=1}^{N}\left(\sum^{n} y\right)_i.$$

Clearly, each unit of the population occurs in n of the N possible circular systematic sample means. Hence,

$$\sum_{i=1}^{N}\left(\sum^{n} y\right)_i = n\sum_{i=1}^{N} Y_i,$$

which on substitution in $E(\bar{y})$ proves the theorem.

What to do when $N \neq nk$

One of the following possible procedures may be adopted when $N \neq nk$

i. Drop one unit at random if sample has $(n+1)$ units.

ii. Eliminate some units so that $N = nk$.

iii. Adopt circular systematic sampling scheme.

iv. Round off the fractional interval k.

Many times, we are interested in measuring a characteristic of a population on several occasions to estimate the trend in time of population means as a time series of the current value of population mean or the value of population mean over several points of time.

When the same population is sampled repeatedly, the opportunities for a flexible sampling scheme are greatly enhanced. For example, on the hth occasion we may have a part of sample that are matched with (common to) the sample at $(h - 1)^{th}$ occasion, parts matching with both $(h - 1)^{th}$ and $(h - 2)^{th}$ occasions, etc.

Such a partial matching is termed as sampling on successive occasions with partial replacement of units or rotation sampling or sampling for a time series.

Notations:

Let *P* be the fixed population with *N* units.

y_t: value of certain dynamic character which changes with time and can be measured for each unit on a number of occasions, $t = 1,2,...n$.

y_{ij}: value of y on j^{th} unit in the population at the ith occasion, $i = 1,2...,h, j = 1,...,N$.

$\bar{Y}_i = \frac{1}{N}\sum_j y_{ij}$: population mean for the ith occasion.

$S_i^2 = \frac{1}{N-1}\sum_{j=1}^{N}(y_{ij} - \bar{Y}_i)^2$: population variance for the ith occasion.

$$S_1^2 = S_2^2 = ... = S^2$$

$$\rho_{ii^*} = \frac{1}{N-1}\sum_{j=1}^{N}(y_{ij} - \bar{Y}_i)(y_{i^*j} - \bar{Y}_{i^*})$$

is the population correlation coefficient between observations occasion i and i^* ($i < i^* = 1, 2,..., h$)

$\rho = \rho_{12}$

s_i^* : sample of size n_i selected at the ith occasion

s_{im}^* : part of s_i^* which is common to (i.e. matched with) s_{i-1}^*

$$s_{im}^* = s_i^* \cap s_{i-1}^*, i = 2,3,...,h \ (s_{1m} = s_{2m}).$$

Note that s_{1m} *and* s_{2m} are of sizes n_1^* *and* n_2^* respectively.

s_{iu}^* : set of units in s_i^* not obtained by the selection in s_{im}^*,

Often $s_{iu}^* = s_{i-1}^{*c} \cap s_i \ (i = 2,...,h)(s_{1u}^* = P - s_{1m}^*)$.

Note that s_{iu}^* is of size $n_i^{**} (= n_i - n_i^*)$.

$\bar{y}_i$ = sample mean of units in ith occasion.

$\bar{y}_i^*$ = sample mean of the units in s_{im}^* on the ith occasion.

$\bar{y}_i^{**}$ = sample mean of units in s_{iu} on the ith occasion.

$\bar{y}_i^{***}$ = sample mean of units in s_{im} on the (i - 1)th occasion, i = 2,3,...,h.

$$(\overline{y}_2^{***} = \overline{y}_1^{*}, \overline{y}_i^{***} \textit{ depends on } \overline{y}_{i-1} \textit{ and } \overline{y}_{i-1}^{**})$$

Sampling on Two Occasions

Assume that $n_i = n$

$$n_i^* = m$$

$$n_i^{**} = u(= n - m), i = 1, 2.$$

Suppose that the sample s_1^* is an SRSWOR from P. The sample

$$s_2^* = s_{2m}^* \cup s_{2u}^*$$

Where s_{2m}^* is an SRSWOR sample of size m from s_1^* and

s_{2u}^* is an SRSWOR sample of size u from $(P - s_1^*)$.

Estimation of Population Mean

Two types of estimators are available for the estimation of population mean:

1. Type 1 estimators: They are obtained by taking a linear combination of estimators obtained from s_{2u}^* *and* s_{2m}^*.
2. Type 2 estimators: They are obtained by considering the best linear combination of sample means.

Type 1 Estimators:

Two estimators are available for estimating $\overline{Y}_2$

i. $t_{2u} = \overline{y}_2^{**}$

With $Var(\overline{y}_2^{**}) = \dfrac{S_2^2}{u} = \dfrac{1}{W_u}(say)$

ii. t_{2m} = linear regression estimate of $\overline{Y}_2$ based on the regression of y_{2j} on y_{1j}

$$\overline{y}_2^* + b(\overline{y}_1 - \overline{y}_1^*)$$

where $b = \dfrac{\sum_{i \in s_{2m}^*} (y_{1j} - \overline{y}_1^*)(y_{2j} - \overline{y}_2^*)}{\sum_{j \in s_{1m}^*} (y_{1j} - \overline{y}_1^*)^2}$ is the sample regression coefficient.

Recall in case of double sampling, we had

$$Var(\hat{\bar{Y}}_{regd}) = S_y^2\left(\frac{1}{n}-\frac{1}{N}\right) - \rho^2 S_y^2\left(\frac{1}{n}-\frac{1}{n^*}\right)$$

$$= \frac{S_y^2}{n} - \rho^2 S_y^2\left(\frac{1}{n}-\frac{1}{n^*}\right)$$

$$= \frac{(1-\rho^2)}{n}S_y^2 + \frac{\rho^2 S_y^2}{n^*} \quad \text{(ignoring term of order } \frac{1}{N}\text{).}$$

So in this case

$$Var(t_{2m}) = \frac{S_2^2(1-\rho^2)}{m} + \frac{\rho^2 S_2^2}{n}$$

$$= \frac{1}{W_m} \text{ (say).}$$

If there are two uncorrelated unbiased estimators of a parameter, then the best linear unbiased estimator of parameter can be obtained by combining them using a linear combination with suitably chosen weights. Now we discuss how to choose weights in such a linear combination of estimators.

Let $\hat{\theta}_1$ *and* $\hat{\theta}_2$ be two uncorrelated and unbiased estimators of θ, i.e., $E(\hat{\theta}_1) = E(\hat{\theta}_2) = \theta$ and $Var(\hat{\theta}_1) = \sigma_1^2, Var(\hat{\theta}_2) = \sigma_2^2, Cov(\hat{\theta}_1, \hat{\theta}_2) = 0$

Consider $\hat{\theta} = \omega\hat{\theta}_1 + (1-\omega)\hat{\theta}_2$ where $0 \le \omega \le 1$ is the weight. Now choose ω such that $Var(\hat{\theta})$ is minimum.

$$Var(\hat{\theta}) = \omega^2\sigma_1^2 + (1-\omega)^2\sigma_2^2$$

$$\frac{\partial Var(\hat{\theta})}{\partial \omega} = 0$$

$$\Rightarrow 2\omega\sigma_1^2 - 2(1-\omega)\sigma_2^2 = 0$$

$$\Rightarrow \omega = \frac{\sigma_2^2}{\sigma_1^2+\sigma_2^2} = \omega^*, \text{ say}$$

$$\left.\frac{\partial^2 Var(\hat{\theta})}{\partial \omega^2}\right|_{\omega=\omega^*} > 0.$$

The minimum variance achieved by $\hat{\theta}$ is

$$Var(\hat{\theta})_{Min} = \omega^{*2}\sigma_1^2 + (1-\omega^*)^2\sigma_2^2$$

$$= \frac{\sigma_2^4\sigma_1^2}{\left(\sigma_1^2+\sigma_2^2\right)^2} + \frac{\sigma_1^4\sigma_2^2}{\left(\sigma_1^2+\sigma_2^2\right)^2}$$

$$= \frac{\sigma_2^2\sigma_1^2}{\sigma_1^2+\sigma} = \frac{1}{\frac{1}{\sigma_2^2}+\frac{1}{\sigma_1^2}}.$$

Now we implement this result in our case.

Consider the linear combination of t_{2u} *and* t_{2m} as

$$\hat{\bar{Y}}_2 = \omega t_{2u} + (1-\omega)t_{2m}$$

where the weights ω are obtained as

$$\omega = \frac{W_u}{W_u + W_m}$$

so that $\hat{\bar{Y}}_2$ is the best combined estimate.

The minimum variance with this choice of ω is

$$Var(\hat{\bar{Y}}_2) = \frac{1}{W_u + W_m} = \frac{S_2^2(n-u\rho^2)}{(n^2-u^2\rho^2)}.$$

For $u = 0$ (complete matching), $Var(\hat{\bar{Y}}_2) = \frac{S_2^2}{n}$.

For $u = n$ (no matching), $Var(\hat{\bar{Y}}_2) = \frac{S_2^2}{n}$.

Type II Estimators

We now consider the minimum variance linear unbiased estimator of $\bar{Y}_2$ under the same sampling scheme as under

Type I estimator

A best linear (linear in terms of observed means) unbiased estimator of $\hat{\bar{Y}}_2$ is of the form

$$\hat{\bar{Y}}_2^* = a\bar{y}_1^{**} + b\bar{y}_1^* + c\bar{y}_2^* + d\bar{y}_2^{**}$$

where constants a,b,c,d and matching fraction $\lambda\left(=\frac{m}{n}=\frac{n-1}{n}\right)$ are to be suitably chosen so as to minimize the variance.

Assume $S_1^2 = S_2^2$.

Now $E(\hat{\bar{Y}}_2^*) = (a+b)\bar{Y}_1 + (c+d)\bar{Y}_2$.

If $\hat{\bar{Y}}_2^*$ has to be an unbiased estimator of $\bar{Y}_2$, i.e.

$$E(\hat{\bar{Y}}_2^*) = \bar{Y}_2,$$

it requires

$$a+b=0$$

$$c+d=1.$$

Since a minimum variance unbiased estimator would be uncorrelated with any unbiased estimator of zero, we must have

$$Cov(\hat{\bar{Y}}_2^*, \bar{y}_1^{**} - \bar{y}_1^*) = 0 \quad (1)$$

$$Cov(\hat{\bar{Y}}_2^*, \bar{y}_2^* - \bar{y}_2^{**}) = 0 \quad (2)$$

Since

$$Cov(\bar{y}_2^*, \bar{y}_1^{**}) = 0 = Cov(\bar{y}_2^*, \bar{y}_2^{**})$$

$$Cov(\bar{y}_2^*, \bar{y}_1^*) = \frac{\rho S^2}{m}$$

$$Cov(\bar{y}_2^{**}, \bar{y}_1^{**}) = Cov(\bar{y}_2^{**}, \bar{y}_2^*) = 0$$

$$Var(\bar{y}_2^*) = \frac{S^2}{m}$$

$$Var(\bar{y}_2^{**}) = \frac{S^2}{u}.$$

Now solving (1) and (2) by neglecting terms of order $\frac{1}{N}$ we have

$$\begin{aligned}
Cov(\hat{\bar{Y}}_2^*, \bar{y}_1^{**} - \bar{y}_1^*) &= Cov(a\bar{y}_1^{**} + b\bar{y}_1^* + c\bar{y}_2^* + d\bar{y}_2^{**}, \bar{y}_1^{**} - \bar{y}_1^*) \\
&= aVar(\bar{y}_1^{**}) + bCov(\bar{y}_1^*, \bar{y}_1^{**}) + cCov(\bar{y}_2^*, \bar{y}_1^{**}) + dCov(\bar{y}_2^{**}, \bar{y}_1^{**}) \\
&\quad - aCov(\bar{y}_1^{**}, \bar{y}_1^*) - bVar(\bar{y}_1^*) - cCov(\bar{y}_1^*, \bar{y}_2^*) - dCov(\bar{y}_2^{**}, \bar{y}_1^*)
\end{aligned}$$

or

$$-a\frac{\rho S^2}{m}+\frac{cS^2}{m}=\frac{(1-c)S^2}{u}. \quad (3)$$

Similarly, from (2), we have

$$Cov(\bar{y}_2^*, \bar{y}_2^* - \bar{y}_2^{**}) = 0$$

$$\Rightarrow -\frac{aS^2}{m}+\frac{c\rho S^2}{m}=\frac{aS^2}{u}. \quad (4)$$

Solving (3) and (4) gives

$$a=\frac{\lambda\mu\rho}{1-\rho^2\mu^2},\quad c=\frac{\lambda}{1-\rho^2\mu^2}$$

$$\text{where } \mu=\frac{u}{n}=1-\lambda,\ \lambda=\frac{n-u}{n}$$

$$b=-a,\ d=1-c.$$

Substituting a, b, c, d, the best linear unbiased estimator of $\bar{Y}_2$ is

$$\hat{\bar{Y}}_2^* = \left[\lambda\mu\rho(\bar{y}_1^{**}-\bar{y}_1^*)+\lambda\bar{y}_2^*+\frac{\mu(1-\rho^2\mu)\bar{y}_2^{**}}{(1-\rho^2\mu^2)}\right].$$

For these values of a and c,

$$Var(\hat{\bar{Y}}_2^*)=\left(\frac{1-\rho^2\mu S^2}{1-\rho^2\mu^2 n}\right)$$

Alternatively, minimize $Var(\hat{\bar{Y}}_2^*)$ with respect to a and c and find optimum values of a and c. Then find the estimator and its variance.

Till now, we used SRSWOR for the two occasions. We now consider unequal probability sampling schemes on two occasions for estimating. We use the same notations as defined in varying probability scheme.

Des Raj Scheme

Let s_1^* be the sample selected by PPSWR from P using x as a size (auxiliary) variable.

Then $p_i = \frac{x_i}{X_{tot}}$ is the size measure of i, where X_{tot} is the population total of auxiliary variable.

$$s_2^* = s_{2m}^* \bigcup s_{2u}^*$$

where s_{2m}^* is an SRSWR(m) from s_1^* and s_{2u}^* is an independent sample selected from P by PPSWR using u draws $(m+u=n)$.

The estimator is

$$\hat{Y}_{2des} = \omega t_{2m} + (1-\omega) t_{2u};\ 0 \le \omega \le 1$$

where

$$t_{2m} = \sum_{j \in s_{2m}^*} \left(\frac{y_{2j} - y_{1j}}{mp_j} \right) + \sum_{j \in s_1^*} \left(\frac{y_{1j}}{np_j} \right)$$

$$t_{2u} = \sum_{j \in s_{2u}^*} \left(\frac{y_{2j}}{up_j} \right).$$

Assuming

$$\sum_{j=1}^{N} P_j \left(\frac{Y_{1j}}{P_j} - Y_1 \right)^2 = \sum_{j=1}^{N} P_j \left(\frac{Y_{2j}}{P_j} - Y_2 \right)^2 = V_0 \text{ (say)}.$$

For the optimum sampling fraction

$$\lambda = \frac{m}{n},$$

$$\operatorname{Var}\left(\bar{Y}_{2des}\right) = \frac{V_0\left(1+\sqrt{2(1-\delta)}\right)}{2n}$$

Where

$$\delta = \frac{\sum_{i=1}^{N} P_i \left(\frac{Y_{1i}}{P_i} - Y_1 \right)\left(\frac{Y_{2i}}{P_i} - Y_2 \right)}{\sigma_{pps}(y_1)\sigma_{pps}(y_2)}$$

$$Var_{pps}(z) = \sum_{i=1}^{N} P_i \left(\frac{Z_i}{P_i} - Z \right)^2 = \sigma_{pps}^2(z)$$

$$Z = \sum_{i=1}^{N} Z_i.$$

Chaudhuri - Arnab Sampling Scheme

Let s_1^* be the sample selected by Midzuno's sampling scheme,

$$s_2^* = s_{2m}^* \bigcup s_{2u}^*$$

where

s_{2m}^* = SRSWOR sample from s_1^*

s_{2u}^* = sample of size u from P by Midzuno's sampling scheme.

Then an estimator of $\overline{Y}$ is

$$\hat{\overline{Y}}_{2ca} = \alpha t_{2m} + (1-\alpha) t_{2u};\ 0 \le \alpha \le 1$$

where

$$t_{2m} = \sum_{j \in s_{2m}^*} \left(\frac{(y_{2j} - y_1)n}{m\pi_j} \right) + \sum_{j \in s_1^*} \left(\frac{y_{1j}}{\pi_j} \right)$$

$$t_{2u} = \sum_{j \in s_{2u}} \frac{y_{2j}}{\pi_j^*}$$

$$\pi_j = np_j$$

$$\pi_j^* = up_j.$$

Similarly other schemes are also there.

Sampling on More than Two Occasions

When there are more than two occasions, one has a large flexibility in using both sampling procedures and estimating the character.

Thus on occasion i

- one may have parts of the sample that are matched with occasion (i - 1)
- parts that are matched with occasion $(i - 2)_j$
- and so on.

One may consider a single multiple regression of all previous matchings on the current occasion.

However, it has been seen that the loss of efficiency incurred by using the information from the latest two or three occasions only is fairly small in many occasions.

Consider the simple sampling design where

$$s_i^* = s_{im}^* \bigcup s_{iu}^*,$$

where

s_{im}^* is a sample by SRSWOR of size m_i from $s_{(i-1)}^*$,

s_{iu}^* is a sample by SRSWOR of size $u_i(= n\text{-}m_i)$ from the units not already sampled.

Assume $n_i = n, S_2^2 = S^2$ for all i.

On the ith occasion, we have therefore two estimators

$$t_{iu} = \bar{y}_i^{**} \text{ with } Var(t_{iu}) = \frac{S^2}{u_i} = \frac{1}{W_{iu}}$$

$$t_{im} = \bar{y}_i^* + b_{(i-1)i}(\hat{\bar{Y}}_{(i-1)} - \bar{y}_i^{***})$$

where $b_{(i-1),i}$ is the regression of y_{ij} given by $b_{(i-1),i} = \dfrac{\sum_{s_i^{**}}\left(y_{(i-1)j} - \bar{y}_i^{***}\right)\left(y_{ij} - \bar{y}_i^*\right)}{\sum_{s_i^{**}}\left(y_{(i-1)j} - \bar{y}_i^{***}\right)^2}$.

$$Var(t_{im}) = \frac{S^2(1-\rho^2)}{m_i} + \rho^2 Var(\hat{\bar{Y}}_{(i-1)}) = \frac{1}{W_{im}}$$

assuming that $\rho_{(i-1),i} = \rho, i = 2,3...,$ and terms of order $\frac{1}{N}$ are negligible.

The expression of $Var(t_{im})$ has been obtained from the variance of regression estimator under double sampling

$$V(\hat{\bar{y}}_{regd}) = S_u^2\left(\frac{1}{n} - \frac{1}{N}\right) - \rho^2 S_y^2\left(\frac{1}{n} - \frac{1}{n^*}\right)$$

$$= \frac{S_y^2(1-\rho^2)}{n} - \frac{\rho^2 S_y^2}{n^*}$$

which is obtained after ignoring the terms of $\frac{1}{N}$ by using m_i for n and replacing $\frac{\rho^2 S_y^2}{n^*}\left(= \beta^2 V\left(\bar{x}^*\right)\right)$ by $\rho^2 Var\left(\hat{\bar{Y}}_{(i-1)}\right)$ since $\beta = \rho$ and S_i^2 is constant. Using weights as the inverse of variance, the best weighted estimator from t_{iu} and t_{im} is

$$\hat{\bar{Y}}_{iu} = \omega_i + t_{iu} + (1-\omega_i) + t_{im}$$

where

$$\omega_i = \frac{W_{iu}}{W_{iu} + W_{im}}$$

Then

$$\text{Var}\left(\hat{\bar{y}}_i\right) = \frac{1}{W_{iu} + W_{im}} = \frac{g_i S^2}{n} \text{ say } i = 1, 2 \ldots (g_1 = 1)$$

Substituting

$$\frac{1}{W_{iu}} = \frac{S^2}{u_i}$$

$$\text{in } \frac{1}{W_{iu} + W_{im}} = \frac{g_i S^2}{n},$$

we have

$$\frac{n}{g_i} = u_i + \frac{1}{\frac{1-\rho^2}{m_i} + \frac{\rho^2 g_{i-1}}{n}}.$$

Now maximize $\frac{n}{g_i}$ with respect to m_i so as to minimize $\text{Var}\left(\hat{\bar{y}}_i\right)$. So differentiate $\frac{n}{g_i}$ with respect to m_i and substituting it to be zero, we get

$$\frac{(1-\rho^2)}{m_i^2} = \left(\frac{1-\rho^2}{m_i} + \frac{\rho^2 g_{i-1}}{n}\right)^2$$

$$\Rightarrow \hat{m}_i = \frac{n\sqrt{1-\rho^2}}{g_{i-1}(1+\sqrt{1-\rho^2})}.$$

Now the optimum sampling fracture $\frac{\hat{m}_i}{n}$ can be determined successively for i=2,3,... for given values of ρ.

Substitute this in the expression of $\frac{n}{g_i}$, we have

$$\frac{1}{g_i} = 1 + \frac{\left(1-\sqrt{1-\rho^2}\right)}{g_{i-1}\rho^2}$$

or

$$q_i = 1 + aq_{i-1}$$

where

$$q_i = \frac{1}{g_i}, q_1 = 1, a = \frac{\left(1-\sqrt{1-\rho^2}\right)}{\left(1+\sqrt{1-\rho^2}\right)}; 0 < a < 1$$

Repeated use of this relation gives

$$q_{i-1} = 1 + aq_{i-2}$$

$$\Rightarrow q_i = 1 + a(1 + aq_{i-1})$$

$$= 1 + a + a^2 q_{i-1}$$

$$q_{i-2} = 1 + aq_{i-3}$$

$$\Rightarrow q_i = 1 + a + a^2(1 + aq_{i-2})$$

$$\vdots$$

$$= \frac{(1-a^i)}{(1-a)} = \frac{1}{1-a} \text{ as } i \to \infty.$$

For sampling an infinite number of times, the limiting variance factor g_∞ is

$$g_\infty = 1 - a = \frac{2\sqrt{1-\rho^2}}{1+\sqrt{1-\rho^2}}.$$

The limiting value of $V\left(\hat{\hat{Y}}_i\right)$ as $i \to \infty$ is

$$\lim_{i\to\infty} Var(\hat{\hat{Y}}_i) = Var(\hat{\hat{Y}}_\infty) = \frac{2S^2\sqrt{1-\rho^2}}{n\left(1+\sqrt{1-\rho^2}\right)}.$$

The limiting value of optimum sampling fraction as $i \to \infty$ is

$$\lim_{i\to\infty} \frac{\hat{m}_i}{n} = \frac{\hat{m}_\infty}{n} = \frac{\sqrt{1-\rho^2}}{g_\infty\left(1+\sqrt{1-\rho^2}\right)} = \frac{1}{2}.$$

Thus for the estimation of current population mean by this procedure, one would not have to match more than 50% of the sample drawn on the last occasion.

Unless ρ is very high, say more than 0.8, the reduction in variance $(1-g_h)$ is only modest.

Type II Estimation

Consider

$$\hat{\bar{Y}}_i = a_i\hat{\bar{Y}}_{i-1} + b_i\bar{y}_{i-1}^{**} + c_i\,\bar{y}_i^{***} + d_i\bar{y}_i^{**} + e_i\bar{y}_i^{*}.$$

Now

$$E(\hat{\bar{Y}}_i) = (a_i + b_i + c_i)\bar{Y}_{(i-1)} + (d_i + e_i)\bar{Y}_i.$$

So for unbiasedness,

$$c_i = -(a_i + b_i)$$
$$d_i = 1 - e_i$$

An unbiased estimator is of the form

$$\hat{\bar{Y}}_i = a_i\hat{\bar{Y}}_{(i-1)} + b_i\bar{y}_i^{**} - (a_i + b_i)\bar{y}_i^{***} + d_i\bar{y}_i^{**} + (1-d_i)\bar{y}_i^{*}$$

To find optimum weights, minimize $V\left(\hat{\bar{Y}}_i\right)$ with respect to a_i, b_i, d_i.

Alternatively, one can consider that

$$\hat{\bar{Y}}_i = a_i\hat{\bar{Y}}_{i-1} + b_i\bar{y}_i^{**} - (a_i + b_i)\bar{y}_i^{***} + d_i\bar{y}_i^{**} + (1-d_i)\bar{y}_i^{*}$$

must be uncorrelated with all unbiased estimators of zero. Thus

$$Cov(\hat{\bar{Y}}_i, \bar{y}_{i-1}^{**} - \bar{y}_i^{***}) = 0$$

$$Cov(\hat{\bar{Y}}_{i-1}, \bar{y}_{i-1}^{**} - \bar{y}_i^{***}) = 0$$

$$Cov(\hat{\bar{Y}}_i, \bar{y}_{i-2}^{**} - \bar{y}_{i-1}^{***}) = 0.$$

Using these restrictions, find the constants and get the estimator.

Permissions

All chapters in this book are published with permission under the Creative Commons Attribution Share Alike License or equivalent. Every chapter published in this book has been scrutinized by our experts. Their significance has been extensively debated. The topics covered herein carry significant information for a comprehensive understanding. They may even be implemented as practical applications or may be referred to as a beginning point for further studies.

We would like to thank the editorial team for lending their expertise to make the book truly unique. They have played a crucial role in the development of this book. Without their invaluable contributions this book wouldn't have been possible. They have made vital efforts to compile up to date information on the varied aspects of this subject to make this book a valuable addition to the collection of many professionals and students.

This book was conceptualized with the vision of imparting up-to-date and integrated information in this field. To ensure the same, a matchless editorial board was set up. Every individual on the board went through rigorous rounds of assessment to prove their worth. After which they invested a large part of their time researching and compiling the most relevant data for our readers.

The editorial board has been involved in producing this book since its inception. They have spent rigorous hours researching and exploring the diverse topics which have resulted in the successful publishing of this book. They have passed on their knowledge of decades through this book. To expedite this challenging task, the publisher supported the team at every step. A small team of assistant editors was also appointed to further simplify the editing procedure and attain best results for the readers.

Apart from the editorial board, the designing team has also invested a significant amount of their time in understanding the subject and creating the most relevant covers. They scrutinized every image to scout for the most suitable representation of the subject and create an appropriate cover for the book.

The publishing team has been an ardent support to the editorial, designing and production team. Their endless efforts to recruit the best for this project, has resulted in the accomplishment of this book. They are a veteran in the field of academics and their pool of knowledge is as vast as their experience in printing. Their expertise and guidance has proved useful at every step. Their uncompromising quality standards have made this book an exceptional effort. Their encouragement from time to time has been an inspiration for everyone.

The publisher and the editorial board hope that this book will prove to be a valuable piece of knowledge for students, practitioners and scholars across the globe.

Index

www.ingramcontent.com/pod-product-compliance
Lightning Source LLC
Chambersburg PA
CBHW061955190326
41458CB00009B/2875

9781682854945